Materials

A **SCIENTIFIC** *Book*
AMERICAN

Materials

W. H. FREEMAN AND COMPANY
San Francisco

Copyright © 1967 by SCIENTIFIC AMERICAN, Inc.
All rights reserved. No part of this book may
be reproduced in any form without permission in
writing from the publisher, except by a
reviewer, who may quote brief passages and
reproduce not more than three illustrations in a
review to be printed in a magazine or newspaper.

The thirteen chapters in this book originally
appeared as articles in the September 1967 issue
of *Scientific American*.

Library of Congress Catalogue Card Number 67-30439
ISBN: 0-7167-0970-8

Printed in the United States of America

9 8 7 6 5

Introduction

To make materials, atoms join together in the solid state. The great diversity of materials reflects essentially the fact that atoms come in different sizes and form bonds of greater or lesser strength and directionality with one another. Depending on their relative sizes and the nature of the bonds that engage them, atoms assemble in more or less tight, stable and regular structures. The atomic architecture of these structures determines the gross properties of each kind of material.

Such understanding has united the specialized knowledge and the empirical lore surrounding the traditionally distinctive families of materials into a single science and technology. High-polymer chemists find themselves talking the same language as metallurgists. A new generation of materials scientists works back from end-use specifications to the choice, adaptation and invention of materials. This development is so recent that it is just now being expressed in the restructuring of curricula in the schools and the reorganization of research and engineering staffs in industry.

The highly structure-dependent mechanical properties of materials dictate the end uses in which the vast bulk of them is employed. For any given set of mechanical requirements, however, the materials engineer today has a range of choice that is not limited by the boundaries of any one of the materials industries. In this volume, that continuity of characteristics and choice lends a unitary theme to the five chapters that describe the commonly distinguished major families of materials. Choice is sharpened at every point by the competitive perfection of mechanical properties. Thus, steels are made to meet such seemingly impossible combinations of requirements as "easily formed" and "resistance to deformation," or "easy machinability" and "high surface hardness." Yet metallurgists, despite the head start gained by the earlier maturation of their science, are beleaguered in every market by the virtuosity of the polymer chemist and the newly won sophistication of the ceramicist. The competition gets its boundary-cutting edge from the research and development carried forward by the end-use industries. With no economic ties to raw-materials sources, their laboratories are among the foremost centers of the new science of materials.

Turning from the highly structure-dependent mechanical properties, the new science has mobilized, with great cunning, the electrical, magnetic, thermal, optical and chemical properties of materials. These depend upon the relative excess, or deficiency, of the electrons that establish structural bonds, and upon their availability and mobility. Between the familiar conductors (excess electrons) and insulators (no free electrons) precise control of structure has created the family of semiconductors (slight excess or deficiency of electrons) and refashioned the technology of electronics. The same precise control has enabled scientists to extend the optical properties of solids to include the stimulated emission of radiant energy in the microwave (maser) and visible light (laser) spectra.

Competition among materials in contemporary U.S. industrial technology is obliterating the familiar landmarks that formerly staked out the markets of the primary-materials industries. Materials, from the point of view of the fabricator, present a continuous range of alternatives: the market, from the point of view of the primary-materials producers, presents a correspondingly diverse range of challenge and opportunity. The connection between buyer and seller depends everywhere increasingly on the collaboration of materials scientists on both sides of the transaction.

The chapters in this book were first published as articles in the September, 1967, issue of *Scientific American*. This was the eighteenth in the series of single-topic issues published annually by the magazine. The editors herewith express appreciation to their colleagues at W. H. Freeman and Company, the book-publishing affiliate of *Scientific American*, for the enterprise that has made the contents of this issue so speedily available in book form.

THE EDITORS*

September, 1967

---

*BOARD OF EDITORS: Gerard Piel (Publisher), Dennis Flanagan (Editor), Francis Bello (Associate Editor), Philip Morrison (Book Editor), Barbara Lovett, John Purcell, James T. Rogers, Armand Schwab, Jr., C. L. Stong, Joseph Wisnovsky, Jerome Snyder (Art Director), Samuel L. Howard (Assistant Art Director)

Contents

Materials 1
CYRIL STANLEY SMITH
A study of materials, with special reference to their basic nature and properties.

The Solid State 21
SIR NEVILL MOTT
Materials can be divided into two general categories: crystalline and amorphous.

The Nature of Metals 39
A. H. COTTRELL
Such properties of metals as luster and conductivity flow from the metallic bond.

The Nature of Ceramics 55
JOHN J. GILMAN
The ionic and covalent bonds render ceramics hard, brittle and resistant to heat.

The Nature of Glasses 69
R. J. CHARLES
Glasses are normally amorphous materials. How do they avoid being crystalline?

The Nature of Polymeric Materials 85
HERMAN F. MARK
The chainlike molecules of polymers form both amorphous and crystalline arrays.

7 ✗ *The Nature of Composite Materials* 97
ANTHONY KELLY
   Two materials can be combined to obtain properties that neither shows by itself.

8 *The Thermal Properties of Materials* 111
JOHN ZIMAN
   Heat is conducted through materials by the particle-like waves known as phonons.

9 ✗ *The Electrical Properties of Materials* 127
HENRY EHRENREICH
   There is a wide spectrum from the best electrical conductor and the best insulator.

10 *The Chemical Properties of Materials* 145
HOWARD REISS
   The chemical reactions that are familiar in liquids also take place within solids.

11 *The Magnetic Properties of Materials* 161
FREDERIC KEFFER
   The magnetism of a strong magnet resides in the magnetism of single electrons.

12 ✗ *The Optical Properties of Materials* 177
ALI JAVAN
   What is the basis for such properties as transparency, opacity and laser action?

13 *The Competition of Materials* 191
W. O. ALEXANDER
   Deeper knowledge of materials gives freer reign to economics in choosing them.

The Authors 201

Bibliography 207

Materials

Vitrarij

Presenting a study of materials in two aspects: (1) the fundamental nature of metals, ceramics, glasses, polymers and composite materials and (2) the properties that all materials possess in varying degrees.

CYRIL STANLEY SMITH

Materials

In acknowledging the importance of materials in human culture (and conceding that earlier times must have been better than the present) the Greeks posited four ages of man—Gold, Silver, Bronze and Heroic—before their own Iron Age. For all that, materials have been pretty much taken for granted, and their diversity and adaptability has rarely been consciously considered. This lack of interest came about, I believe, mainly because man discovered at an early period a variety of balanced materials that were available without further study for a wide range of uses. The designer simply did not think of applications that required more than stone, fired clay, cement, the common biological materials and the alloys of the ancient seven metals. Moreover, although the philosophy and science of matter is virtually coeval with philosophy and science in general, the science of materials is quite recent. The most useful properties are the structure-sensitive ones with which the classical physicist was utterly incapable of dealing and therefore did not consider to be physics.

Glass is being made in the facing illustration from Georgius Agricola's De Re Metallica, *published in Germany in 1556. The men around the beehive-shaped furnace are holding blowpipes for blowing the glass.*

In a way it is the merging of two opposing schools of Greek philosophy of matter that gave rise to the modern science of materials. The elemental qualities of Aristotle are seen to arise in the geometric forms of the Pythagoreans. Now that the ultimate structure and gross properties can be related fundamentally, it is seen that there is less difference between different kinds of materials than had been supposed when they were the basis of totally separate crafts and industries. Properties can now be designed, not just selected. With each new design, however, the materials scientist and the engineer must still seek the optimum compromise between workability in the shop and durability in use.

One of the great stages in man's development was the discovery that he could change the very nature of materials. It must have given early man an almost godlike feeling of control over nature as he turned clay to stone by heating it. The nearly unlimited formability in one state is replaced by high stability in the other because of changes in the complicated relation (on both the atomic scale and a larger one) between the microcrystals of the clay substance and water, and the chemical change and complete recrystallization that occurs on firing above a red heat.

Ceramics, the earliest inorganic materials to be structurally modified by man, exemplify the diversity of structure-related properties better than any other. Except for electrical and magnetic effects, almost all the properties of solids that are the concern of modern solid-state physicists were exploited by early ceramists. Shaping involved moisture-dependent plasticity and thixotropy; decorative textures were derived from vitrification and devitrification, the nucleation of various gaseous and crystalline phases, and local variations of expansivity, viscosity and surface tension; colors depended on various states of oxidation, on abnormal ionic states, on excitons and on structural imperfections in crystals. All these phenomena were discovered for sheer aesthetic pleasure, not utility.

The earliest glazing technique is in many ways the most interesting. This was a Sumerian invention made famous after 4000 B.C. as Egyptian blue faience. It was not, like later glazes, a melted premix of glass-forming materials but was made in a kind of cementation process during which potash was drawn by capillarity to react with the surface of a preformed body of siliceous particles to form a glassy coat of copper-colored eutectic silicate that did not "wet" the powder in which it was embedded. The process is still in use in Iran, where it was recently discovered by Hans E. Wulff of the University of New South Wales. Its implications for the history of sand-cored glass-vessel technology and in various metallurgical cementation processes have yet to be explored.

The earliest use of metal followed not long after the first wide use of fire-hardened clay. Metals owe their main utility to the fact that they are rigid below a certain stress (high enough for most service uses) but become quite plastic above a certain stress (low enough to be surpassed locally by the concentrated action of a tool). Above a certain temperature they become liquid and can be shaped by casting.

Metals began, as did ceramics and many other technological innovations, somewhere in the region that today comprises eastern Turkey, northern Iraq and northwestern Iran. Hammered copper objects in the form of dress ornaments, necklace beads and the like are known from the eighth millennium B.C. They are made of unmelted, unannealed native copper, simply hammered and cut to shape. Native copper seems to have been used in the Middle East mainly for decorative purposes, and no large pieces were fabricated. Conversely, the artisans of the Old Copper Culture in the Great Lakes region of North America (3000–1000 B.C.) hammered sizable pieces of native metal into knives, spears and agricultural implements. The microstructure of these objects shows that the metal had been worked hot or frequently annealed at a red heat in the course of the shaping operation.

In the Middle East cold-hammered native metal was largely displaced around 4000 B.C. by copper smelted from ores. Because copper ores are far more abundant than the native metal, this was a development of vast economic significance. Alloying soon followed. The first copper alloys came from the smelting of ores that contained arsenic and antimony as naturally occurring impurities. These alloys not only are harder than pure copper but also have a lower melting point and give castings with a much better surface and internal soundness. About 3000 B.C. came the bronzes. For two millenniums thereafter the alloys of copper and tin dominated metallurgy in spite of the scarcity of the ores of tin; they combined ease of shaping, beauty and general serviceability better than any other metallic material.

Melting was probably first used to convert copper scrap into lumps that could be hammered into shape, then for casting shaped objects directly. Artisans first used open molds of clay or stone, then molds in two or more pieces assembled to produce objects with more complicated shapes, and eventually clay molds made around a wax pattern that could be burned out, and so removed most restraints in design.

By the time bronze was discovered, elaborate methods of shaping and joining the precious metals—gold and silver—had appeared. Jewelry from the royal graves at Ur in Mesopotamia (around 2600 B.C.) exhibits higher standards, both aesthetic and technical, than most objects made today, and shows that their makers could reproducibly exploit most of the properties of metals that only now are being scientifically explained.

It was economics rather than the inherent qualities of the metal that gave rise to the Iron Age. Unless iron is alloyed with carbon to make steel and is then hardened by heat treatment (for some curious reason it was rarely hardened by cold work) it is weaker and in every way inferior to cold-worked bronze. It is also more difficult to make and more variable in its properties. But iron ores are far more common, and 1,000 inferior iron swords would outweigh 10 good swords of bronze.

Once steel could be reproducibly heat-treated, of course it became supreme. The early metallurgy of iron was quite different from that of copper. Pure iron was unmeltable in any furnace available prior to the 19th century. If iron is heated in a fire long enough to absorb carbon from the charcoal fuel, it changes its properties greatly, becoming first steel, then cast iron, which is not much more difficult to melt than copper. The cast product, however, is relatively brittle. Before A.D. 1400 it was little appreciated except in China, where it was early used for making agricultural tools, stoves and minor works of art. Farther west iron remained for millenniums the smith's metal. The iron sponge resulting from the reduction of a pure ore with charcoal in a hearth was unmeltable, but it was malleable and easily consolidated by hammering when white-hot. Shaping was done entirely by the skillful use of the hammer. Such "wrought iron" always contained residual inclusions of slag and rocky matter that greatly weakened it.

Steel was easy to discover, but the recognition that it was an alloy was long delayed. Unlike bronze, which was consciously made by mixing things, the alloying of carbon with iron occurred invisibly as it was absorbed incidentally to the processing of the metal. The all-important presence of carbon in steel and cast iron was not discovered until 1774, the same year the chemical role of atmospheric oxygen was discovered. Before this both philosophers and blacksmiths had thought that steel was a purer form of iron—not illogically, because iron turned to steel after prolonged heating in a fire that did purify most things.

Heat treatment is, after work-hardening and alloying, the third basic way of modifying the properties of a metal. Heating most nonferrous metals removes the strain-hardening that results from cold work and causes recrystallization, which puts the metal into a soft state ready for further deformation. In steel the effect is more drastic. At a certain critical "transition" temperature the crystal form of iron changes, and carbon can go into solid solution in the crystal lattice of the iron. On cooling iron rapidly, as by quenching in cold water, an intensely hard metastable phase results.

The date by which men had learned to control the heat treatment of steel reasonably well is still uncertain. The best of the ancient smiths were those of Luristan, in the ninth and eighth centuries B.C. Yet their beautifully forged swords have a microstructure clearly showing that the metal, usually fairly high in carbon, was hot-worked at a temperature indiscriminately above and below the transition temperature and was not quenched. The weapons of the Greeks and the Romans were sometimes quench-hardened, but rarely and under poor control. To achieve the desired temper directly by quenching was a critical and difficult operation, particularly when failure could as often be due to a wrong carbon content as to a wrong cooling rate. There was no way to measure either. The modern two-stage process—quenching followed by low-temperature reheating to let down the hardness to an appropriate degree—is much easier to control. It is small wonder that the few sword blades of antiquity that were perfectly heat-treated became legendary.

Yet the early Near Eastern empirical discoverers of materials did their work well. They found nearly all alloy compositions that could have been made by charcoal reduction from recognizable natural minerals and that were fit for general service. The tools, guns, gadgets and cathedrals of the Middle Ages, the instruments for the rise of modern science, the machines and structures of the 19th-century engineers—all were made of materials that had been known centuries before the rise of Greece. Metallurgists had been busy in the interim, increasing the scale and the economy of production and the reliability of the products, but neither they nor the user-smiths felt the need for new compositions. Development much beyond the old required new attitudes, new knowledge and new needs, all of which came together not much before the beginning of the 20th century.

The earliest recorded speculations on the nature of matter are those of the Greek philosophers, whose writings reflect appreciable contact with the artisan's knowledge of the behavior of matter. The atomism of Democritus was a natural—almost inevitable—deduction from an examination of the graded textures of stones and ceramics and from the obvious relation between the properties of bronze and steel variously treated and the texture on the fractured surface of a broken piece. Aristotle could specify in a litany of 18 opposites the qualities that a craftsman would observe and exploit: meltable or unmeltable, viscous or friable, combustible or incombustible and so on. The three familiar states of the aggregation of matter and their relation to energy gave him his four elements: earth, water, air and fire. This was, in a way, good physics, but the "chemistry" that arose in the attempt to account for various substances by the combination of the associated qualities ended in nonsense.

The alchemists sought a relation between the qualities of matter and the principles of the universe. One of their goals—transmutation—was to change the association of qualities in natural bodies. In the days before the chemical elements had been identified it was a perfectly sensible aim. What more proof of the validity of transmutation does one need than the change in the quality of steel reproducibly effected by fire and water? Or the transmutation of ash and sand into a brilliant glass gem, or mud into a glorious Attic vase or Sung celadon pot? Or the conversion of copper into golden brass? Of course today we know that it is impossible to duplicate simultaneously all the properties of gold in the absence of atomic nuclei having a positive charge 79. One way to secure a desired property is still to select the chemical entities involved, but much can also be done by changing the structure of the substance. Modern alchemy is more solid-state physics than it is chemistry, but it could not have appeared until chemists had unraveled the nature and number of the elements.

In the 16th century the principles of Paracelsus—salt, sulfur and mercury—displaced the elemental qualities of Aristotle, expressing an intuitive awareness of the distinctly different properties associated with ionic, molecular and metallic bonds, but this was soon displaced by a chemistry based solely on analytically determined composition. For two centuries chemists were relatively uninterested in properties. The overthrow of caloric and phlogiston in the 18th century represented a similar turning away from physical concepts to a more precise chemistry. Chemists were rightly excited by the demonstration that the reduction of copper or iron from these ores represented the subtraction of oxygen, not the addition of phlogiston. Nevertheless, in today's quantum theory of solids the reduction of a metal can be considered as the boosting of a valence electron into the conduction band. The electron is a small thing, as insubstantial as phlogiston itself, yet it is the return of the electron from the oxygen ion to the metal that confers all the properties that make metals interesting or useful. In a way the phlogistonists were right, and the very real advance that came with the chemical revolution was paid for by the loss of an important viewpoint.

In physics the situation was somewhat the same. The great advances of the 17th century concerned mainly mass and other aspects of matter that were not sensitive to structure. The corpuscular philosophers, both the Cartesians and their rival atomists, spelled out explanations of plasticity, ductility and strength based on *ad hoc* assumptions about the manner in which particles were shaped and stacked in contact with one another, to slide, distort and change neighbors under mechanical stress or change in the chemical environment. These were ingenious and often right (if the corpuscles are appropriately and flexibly interpreted) but quite incalculable. Such corpuscular philosophical speculation vanished under the impact of Newton's mathematical methods, and physics could not seriously return to questions about the solid state until our own time.

The first three significant printed books on materials summarized a substantial accumulation of empirical knowledge. Vannoccio Biringuccio's *Pirotechnia* (1540) presents a comprehensive account of the operations of the foundryman and the smith in alloying and shaping materials for diverse end uses. Georgius Agricola's *De Re Metallica* (1556) gives superb detail on the mining of ore and smelting of ores, with disproportionate emphasis on the nonferrous metals and near neglect of iron. Lazarus Ercker's book of 1574 (with the resounding title *Beschreibung Allenfurnemisten Mineralischen Ertzt unnd Berckwercksarten*) presents the quantitative laboratory approach of the assayer.

These three approaches to metals—production, utilization and analysis—have moved forward but have remained somewhat distinct ever since. For geological and economic reasons the production of metals has been conducted on a large scale, whereas end-product fabrication was the work of multifarious smiths and small enterprises. Something approaching professional metallurgy developed, therefore, first in the service of the princely capitalists or others who controlled the large mines and smelters.

The production of materials other than the metals contributed rather little to the growth of their science. Producers of stone, cement and wood operated on too small and local a scale. For a while, early in the 18th century, the desire of European potters to duplicate Oriental ceramics gave drive to the beginnings of analytical chemistry, but thereafter the science of ceramics lagged; the successful ceramic body had too fine a structure to be seen microscopically and was too complex in composition to yield to simple analysis.

Analytical chemistry owes its origins largely to the methods of the assayer. Economics inspired him to develop ways to recover trace amounts of gold from ores and alloys. He did this quantitatively, exploiting his knowledge that the mass and identity of the metals were preserved through many stages of solution, partition

Bellows are worked at a large smith's hearth in a woodcut from Vannoccio Biringuccio's Pirotechnia. *This book, which was published in Italy in 1540, was a pioneering work on materials.*

and precipitation in a variety of solvents, mostly nonaqueous ones. The first table of chemical affinity, laid out by Etienne Francois Geoffroy in 1718, was essentially a putting in order of the separatory reactions long known to the assayer. Assayers knew a lot about oxides, although they did not know oxygen. They had measured the increase in weight that occurred when lead was turned into litharge to leave behind the shiny bead of gold or silver on their cupels, but to see the theoretical significance of this needed a different kind of curiosity.

The structural side of the science of materials, in spite of its early start in fracture tests and corpuscular philosophy, was slower to blossom. In 1722 Réné Ferchault de Réaumur published an outstanding work on iron, based on observed and hypothetical changes of structure on the level that today we associate with the microstructure. In the best scientific tradition he designed laboratory experiments aimed at checking and improving the theory, and from these he developed an important industrial material, malleable cast iron. His work came, however, at the very end of the period during which Cartesian corpuscular theories could be taken seriously by scientists. Newtonian rigor displaced this kind of structural speculation; microcrystalline grains came back into science only at the end of the 19th century, following the discoveries of the microstructure of steel by Henry Clifton Sorby in 1864.

Mineralogists meanwhile had been studying the symmetries of the external shapes of crystals, and the mathematics of the crystal lattice had been developed without the nature of the units being generally recognized: Johannes Kepler's and Robert Hooke's insight showing how the stacking of spheres in contact gave rise to crystalline polyhedrons was forgotten, and the unit had become prismatic boxes in which molecules were placed. Molecular composition rather than crystal structure was the basis of almost all 19th-century physical and chemical discussions of solids. The dominance of the molecule (so supremely justified in organic chemistry and in the kinetic theory of gases) was replaced by today's structural-atomistic viewpoint largely as a result of a new experimental technique.

In 1912 X-ray diffraction was discovered and soon applied to the study of the structure of solids by Lawrence Bragg and his followers. It at once gave a measurable physical meaning to structure on an atomic scale, and made this as real as the larger-scale structures that had been revealed by Sorby's microscopic methods half a century earlier. It was a physicist's method par excellence, and a fundamental one, which served to relate much of the unconnected data of the chemist and metallurgist.

For a time the X-ray-diffraction results led to the construction of too idealized a picture. Then the role of imperfections was perceived, first chemical, then electrical, then mechanical errors in the building of crystals. The last served to explain the deformability of metals as well as the nature of the interface between crystal grains, the old grain boundary about which practical metallurgists had long speculated because of its great practical importance.

Although still dominant, metals thereafter lost their unique position in scientific studies of materials. Ceramics combined all the interesting crystalline complexity of metals with the electrical interest of semiconductors. Organic chemistry had been developing rapidly in the 19th century as analytical methods became available. The awareness that many compounds with the same composition have different properties engendered the organic chemist's particularly fertile concept of structure. Molecular architecture began almost as a notational device but soon became a central part of organic chemistry and was ready to join with X-ray crystallography in guiding the development of the complicated structures that endow synthetic polymers with their properties.

Foremost in the new understanding of materials is the realization that the properties of all types of material arise from their structure, from the manner in which their constituent atoms aggregate into hierarchies of molecular or crystalline order or into disordered amorphous structures. Moreover, the properties of bulk matter of all kinds depend strongly on the structure of imperfections, either purely architectural or chemical, in the main array. Most of the properties observed and exploited in materials are cooperative properties of the aggregate rather than of the constituent atoms and simple molecules that had perforce been overemphasized by 19th-century investigators. It is the arrangements of the outer electrons of the atoms that are of prime importance, and these are strongly modified by the configuration of neighboring ones.

The 93 species of stable atoms, or even the 10 most common ones in the earth's crust, allow for almost innumerable combinations. All materials depend on the five types of bonding in solid-state physics [see "The Solid State," by Sir Nevill Mott, page 21]. Each bond type is in essence a specific structural framework for electron interaction. The commonest structural materials depend either on the behavior of the valence electron in a relatively free state in a crystal (which accounts for all metals with a close-packed crystal structure), on the unique behavior of the hydrogen atom (so important in many organic materials) or on the rather special properties of two atoms that form highly directed bonds in the tetrahedral configuration recently popularized by Buckminster Fuller as the simplest basis for stable three-dimensional structures. From the carbon atom come all organic compounds, in which atoms are tightly linked to their neighbors in close configuration to provide polyhedral or linear molecules that themselves can be bonded in ever larger hierarchies. The silicon atom, with a similar outer-electronic configuration, gives rise to the silicon dioxide (SiO_2) tetrahedron that is the basis of all silicates and hence the majority of stones, cements, glasses and ceramics. It is indeed a fortunate circumstance that these two elements,

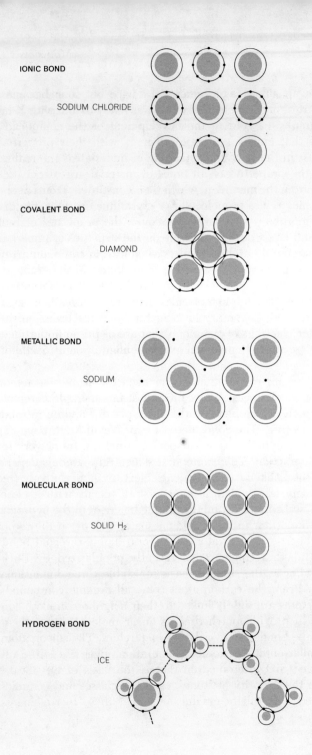

which happen to be present on the earth in great abundance because of their nuclear and geochemical properties, should have the configurational requirements of their outer electrons such that they give rise to such interesting effects.

If today there is a distinguishable materials science and engineering, it is partly owing to the users' preoccupation with physical properties rather than chemical constitution. But it is even more owing to the fact that there is unity in the rules by which atoms and molecules join together in groups and groups of groups. Once the principles of interaction and the possible aggregative geometries have been outlined they will seem to be almost universally applicable. This understanding is not only effective in guiding the development of useful materials; it is also curiously satisfying to the scientist, because it makes him at home in strange landscapes. Of all approaches to nature, a science based on structure seems most able to unite the microscopic and the macroscopic, theory and practice, intuition and logic, beauty and utility. Perhaps it was not chance that involved aesthetics in the early discovery of materials.

Materials, of course, have been as necessary in the scientific laboratory as in the studio or workshop. Those with which to build the first scientific instruments were ready to hand. Astronomers found their materials for telescopes, even achromatic ones, in common brass and in the crown and flint glasses developed for domestic use. There was little need to improve on the magnetism of iron, steel and the lodestone until the electrical industry developed. However, many properties now of the greatest importance

Five types of bond that hold all materials together are shown schematically. In the ionic bond the atoms have either lost an electron or gained one, so that their outer electron shell is complete. Thus they cannot share electrons, but since they are electrically charged by virtue of having gained or lost an electron they are attracted to atoms of the opposite charge. In the covalent bond pairs of atoms share their outer electrons in filling their outer shells. In the metallic bond all the atoms share all the valence electrons. The molecular bond, also known as the van der Waals bond, arises from the displacement of charge within electrically neutral atoms or molecules, which produces a weak attraction between them as they approach each other. The hydrogen bond, also a weak one, is mediated by the hydrogen atom and is possible only because of the atom's small size and the ease with which its charge can be displaced. All these bonds are idealized; most materials involve some combination of them.

were discovered only through scientific research. The whole approach to materials began to change as a result.

In developing improved electrical and magnetic properties the laboratory scientist far outpaced the artisan. The only practical precursors to the discovery of voltaic electricity were some electrochemical replacement reactions (the assayer's recovery of silver from spent parting acids, and the large-scale winning of copper from mine waters with iron) and the observation of the accelerated corrosion of iron rudder brackets on ships sheathed with lead. A great impetus to the study of electrical properties of materials came in 1857, when it was found that some good Spanish copper used in the transatlantic telegraph cable had a conductivity only 14 percent that of the best copper available. Thereafter measurements of resistivity became one of the principal tools in the study of materials of all types. The later discovery of the electron not only led to theories of conductivity, but also when seen in the light of quantum mechanics it became the very cause of the aggregation of matter and the basis of all its properties.

Every use of materials, however trivial, involves selection. It is possible through understanding or experiment to maximize any one property but—and this cannot be overemphasized—in no application is it possible to select a material for one property alone. It is precisely in the balance of one factor against another that the materials engineer finds his challenge and his satisfaction. He must produce a material with the desired compromise of properties, and the scientist must tell how to achieve the structure that gives it. In a way the qualities of Aristotle have returned to be the center of attention, but they are now properties to be desired and designed, not elements to be combined.

Mechanical properties have been the main criterion for the selection of materials from the first use of sticks, and they still limit the size of a building and the cost of an automobile. To understand the basis of strength and ductility was long the main aim of the scientific metallurgist. For centuries steel has provided the standard of strength. Development of heat-treatable alloy steels that enable high strengths to be achieved in large sections coincided with their need in the first automobiles, and incited a new wave of metallurgical research that is not yet over.

In recent years an increasing diversity of materials has come into competition with steel in satisfaction of mechanical requirements set by the entire framework of the economy of production and use. The electric-power and communications industries in particular have set up a constant pressure of innovative demand. It is not only that designers in these industries are interested in a whole new range of electrical, magnetic and optical properties formerly only of laboratory interest; they have also approached their tasks with little ingrained bias toward one or another class of materials in service. Unlike traditional metallurgists, who perforce concentrated on the perfection of one material, they were ready to see all materials in the new conceptual framework of materials science.

That approach to mechanical properties is evidenced in the chapters that describe the major present classes of materials in this book. It is in itself a compelling indication of recent progress in materials science and technology that these chapters should be five in number, because two of them deal with kinds of material hardly known before the beginning of this century. Yet Herman Mark shows how the designers of high polymers are now mounting campaigns of research and development on two major markets of the steelmaker, the construction and automobile industries. And Anthony Kelly describes composite materials that harness the high strengths of perfect crystals found in metals and ceramic "whiskers."

The realization of mechanical properties is perhaps the central line of materials development. It is wrong, however, to overemphasize it, because other properties have been equally improved once the need for them was recognized. As the chapters in this book devoted to these properties show, each of them may be sought and developed effectively in more than one of the major families of materials.

Excellent electrical conductors were found even in the days of frictional electricity by testing the common metals, but years of development were necessary to achieve the oxidation resistance needed for high-temperature electrical service. (The nickel-chromium alloys so developed contributed to the development of high-temperature materials for mechanical service.) The development of ductile tungsten filaments needed for the incandescent

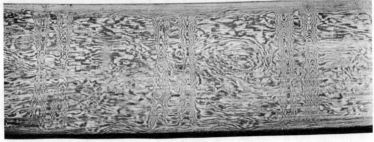

*Four sword blades in New York's Metropolitan Museum of Art reflect three stratagems used by medieval smiths to improve the properties of iron. At top left is an eighth-century-*A.D. *Carolingian blade. The laminations running down the middle arise from a complex hammer-welding and forging operation that served to break up slag inclusions and give a composite metal that was both tough and decorative. At top right is a Turkish "Damascus" blade. It was forged from a cast high-carbon steel ingot that had been slowly cooled to precipitate iron carbide in a form that remains visible after intensive forging. The curly pattern on the surface is produced by local variations in deforma-*

lamp led in turn to more fruitful studies of grain-boundary behavior than had any scientifically inspired research beforehand. Alloying invariably increases the resistance of a metal at ordinary temperatures; superconductivity has changed all of this and has opened a whole new field of compounds.

The discovery of a wide range of materials for "solid state" electronic devices involved a fine mixture of technology and science. First the scientific but empirical discovery of the photoconductivity of selenium, then the electric lamp to exploit the properties of carbon, then the early crystal-and-whisker radio sets

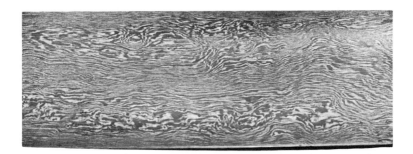

tion. At bottom left is a Persian scimitar, also made of forged high-carbon steel. The regular vertical markings were made by local hammering that distorted the grain of the metal. At bottom right is a Japanese blade. Part of its pattern is the forged-in grain of mechanically heterogeneous metal. The whitish areas are the result of differential heat treatment. The blade was coated with a refractory material that was locally removed to expose the blade and other parts that were to be hardened; the whole blade was then heated and quenched in water. This procedure was unique with Japanese blades, which excel all others.

and at last radar to bring out that prince of semiconductors, silicon. This was followed by the invention of the transistor, which forced the development of both better theory and better practical schemes for producing materials of fantastic purity and controlled impurity. Only slightly less glamorous is the story of the development of new magnetic materials, both "hard" and "soft."

Although every engineering decision adds up to an economic choice, economics often enters on its own as an independent variable. An attractive surface quality may disproportionately increase sales; a slightly greater working stress may win an entirely new market. Even the old property of strength is no longer evaluated straightforwardly; instead of overdesign to ensure permanently against failure, design is now for an expected life.

The demand for different kinds of performance by the various parts of a device or structure is usually met by the use of members made of different materials, each doing the best for its local function: a hard steel journal and a bronze bushing in a wooden machine, paint on the surface for corrosion resistance, drawn and heat-treated steel wire combined with a cast-steel shoe in a suspension bridge, a resistor welded to a conductor, a wooden handle on a silver teapot. Some of the more interesting ancient objects, however, are those in which the properties are made to vary from one part to the other within a single continuous mass. Glazed ceramics provide the earliest examples, but the most notable are steel objects with a variable carbon content and variable hardness in different parts. In the old Japanese sword differential heat treatment produced extreme hardness on the cutting edge, backed, by means of a graded interface, by the toughness and mass of the unhardened steel in the body of the blade.

Damascus steel had an overall heterogeneity on a scale just visible to the unaided eye. All materials have some microscopic or submicroscopic heterogeneity that is more or less natural to them. An exciting new field is the production of composite materials with synthetic heterogeneity.

Without chemical stability in the environment no other property of a material can be utilized. From the first use of gourds and clay pots as cooking vessels to the latest reactor of stabilized stainless steel, the chemical industry has grown symbiotically with the discovery of new corrosion-resistant materials. These are effectively homogeneous on the scale at which they are used. Perhaps the next stage will be to make large members or entire structures with heterogeneity locally responsive to the different service requirements. The engineer will design the local microstructure of the material at the same time that he designs the machine; the whole will be intricately interwoven with no sharp interface between the corrosion-resistant surface and the stress-resisting inside, with variations of elastic modulus between the ends of the beams and the middle, with gradual transitions from parts made of material so that they can easily be welded to others designed for maximum strength. Electrical circuits designed and assembled from separate units are already being challenged by semiconductor microcircuits in which the resistor, capacitor, rectifier and magnet are hardly distinguishable from one another. Such circuits are at present hard to design but not difficult to make, and they are very easy to use.

The habits of thought engendered by these developments may prove to be more important than the devices themselves. It is the limitations of our past unitary scientific theory that has forced the conceptual separation of functions. There is no reason why this should always be so. In a biological organism it is hardly possible to separate the materials of structural function from the materials of electrical and chemical ones. In such an approach lies, I think, the advancing forefront of materials science in the future.

There is a continuous hierarchy of structures and interactions. Nucleons, nuclei, atoms, molecules, crystals, cells, rocks, plants and animals, societies, mountains, continents, planets, stars, universes. Each has its recognizable structure. For study we divide them down into simple categories, one scale at a time, yet the validity of all lies in the fact that there is interplay between scales. As desirable properties are seen to come from the structure as well as from the unit, new methods of thinking will inevitably develop. In the past almost all advance has been in terms of analysis to find units that could be exactly specified; synthesis has been rudimentary. The alchemist's crucible with its complex compositions of matter and changes of structure and qualities was indeed to some degree a symbolic model of the complex structure and action of the universe. If the principles of hierarchial structure can be adequately developed to aid the science of materials, they may in some degree apply also to biological organisms and even to human society.

Materials are solids, and solids are divided into two general categories: crystalline, in which the atoms are stacked in more or less regular arrays, and amorphous, in which they are not.

SIR NEVILL MOTT

The solid state

If you take a paper clip and bend it, it stays bent; it doesn't spring back and it doesn't break. The metal of which the clip is made is said to be ductile. Some other materials are not ductile at all. If you try to bend a glass rod (unless you are holding it in a flame), it will simply break. It is said to be brittle. In this respect, as in many others, glass behaves quite differently from a metal. The difference must lie either in the particular atoms of which metals and glass are made up or in the way they are put together—probably both. There are of course many other differences between metals and glass. Metals conduct electricity and are therefore used for electrical transmission lines; glass hardly conducts electricity at all and can serve as an insulator. Glass is

Etch pits in cadmium sulfide, a semiconductor widely used in photocells, are shaped like hexagons, reflecting the geometric pattern of the atoms that compose the material. In this photomicrograph the etched surface of a single crystal is viewed along the hexagonal axis. The concentric hexagons provide a contour map of the pyramidal pits, which mark defects in the crystalline structure of the material. The photomicrograph was made by Carl E. Bleil and Harry W. Sturner of the General Motors Research Laboratories. The magnification is approximately 475 diameters.

transparent and is used in windows, whereas a sheet of metal more than a millionth of an inch thick is quite opaque.

Students of such matters naturally want to understand the reasons for these differences in behavior. They want to study in detail the mechanical, electrical and optical properties of every kind of solid, and many other properties as well. Moreover, they want to provide the basis for choosing materials with desired properties in every branch of technology. During the past 20 years studies of this kind have been called solid-state physics, or sometimes, since the subject includes a great deal of chemistry, just "solid state." It is a major branch of science that has revealed new and previously unsuspected properties in materials. An example is the properties of semiconductors, knowledge of which has given rise to a flood of technological devices such as the transistor. Indeed, solid-state physics has become one of the most important branches of technology. Today engineers freely use expressions such as "valence band" and "conduction band," which are terms of quantum mechanics as it is applied to solids. In solid state perhaps more than anywhere else quantum mechanics has ceased to be restricted to pure science and has become a working tool of technology.

Of course, solids were the subject of experimental investigation long before quantum mechanics was invented. I shall begin with the fact—known since the earliest studies of electric currents—that metals conduct electricity well and most other materials do not. With the discovery of the electron at the beginning of this century and the realization that it was a universal constituent of matter, it was assumed that in metals some or all of the atoms had lost an electron and that in insulators such as glass they had not. The electrons in a metal were thus free to move about and conduct electricity, whereas the electrons in an insulator were not.

Why this happened in metals had to await the discovery of quantum mechanics, and even now the answer is not quite clear. It has been known for some time, however, how to find the number of free electrons in a metal. The simplest way is based on the Hall effect: in the vicinity of a magnet the electrons carrying a current in a wire are pushed sideways, so that a voltage—the Hall voltage—is set up across the wire. This voltage can be measured, and since it depends only on the speed with which each electron is moving down the wire, whereas the current depends both on the speed and on how many electrons there are, the measurement of both Hall voltage and current enables us to estimate the number of free electrons in a wire. It turns out that in a good conductor such as copper each atom has lost just about one electron. There must be in the metal a very dense gas of electrons, more than 10^{22} of them in a cubic centimeter.

The next question is: How are the atoms themselves arranged? Since the introduction of X-ray crystallography by William Bragg and his son Lawrence in 1913, this has been known for the simpler materials. Solids can be divided into two classes: crystalline and amorphous. In the crystalline group, which is the largest and includes the metals and most minerals, the atoms are arranged in a regular way; in many metals (for instance copper and nickel) they are packed together just as one would pack tennis balls into a box if one wanted to squash in as many as possible. In other metals (for instance iron) the structure is called body-centered cubic; there are four atoms at the corners of a cube and one in the center. The arrangement of atoms in all crystalline solids falls into 14 such categories.

The commonest of the amorphous group of solids is glass. Its atoms are put together in a more disordered way than those of a metal [*see illustration on page 25*]. The structure of an amorphous material is much more difficult to discover than that of a crystalline solid, and considerable effort is now being made to learn more about the arrangement of atoms in such materials.

The crystalline structure of a metal such as iron presents to the eye a formidable array of atoms. How does an electric current manage to flow through such a material? One would think that no electron could get farther than from one atom to another without a collision, and that the electrons must percolate through the crystal the way hailstones sift through the leaves and branches of a tree and fall on someone taking shelter below. If this were the case, it would mean that the rate of drift of an electron gas, and therefore the current for a given voltage, would depend little on whether the arrangement of the atoms was regular or haphazard. That is far from being the truth. One of the most marked characteristics of metals is that they conduct much better at low temperatures than at high ones. For example, the amount of energy

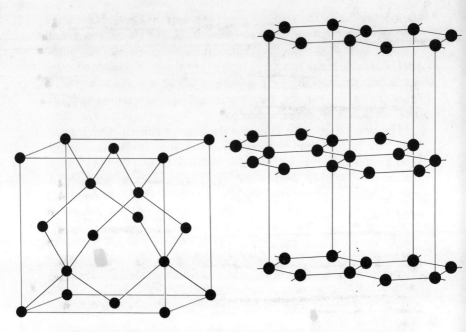

Two forms of carbon exhibit markedly different properties owing to their different crystal structure. Diamond (left) consists of pairs of carbon atoms in a face-centered-cubic array. Each carbon atom is bound to four others. This tightly joined lattice contributes to diamond's hardness. In graphite (right), a soft material, carbon atoms are arranged in layers that are bound by weaker forces.

wasted by resistance in an electric cable is about 10 percent less in a typical Temperate Zone winter than in summer. At the very low temperatures obtainable in cryogenic laboratories an electron can go straight through millions of planes of atoms without being deflected from its path.

Electrical resistance occurs only if the atoms are *not* in a regular array. One such irregularity arises as the temperature increases; the atoms then start to swing around their average positions, each one being displaced up to as much as 10 percent from its normal position. Other evidence is provided by the fact that most metals conduct worse when they melt, and also by the fact that an alloy such as brass (a mixture of copper and zinc) conducts much

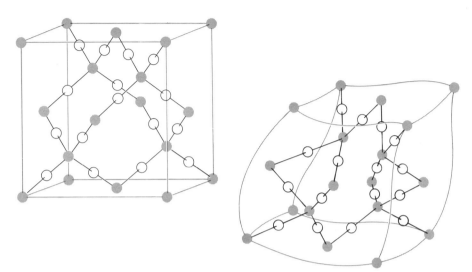

Two forms of silica demonstrate how molecules of the same composition can exist either as a crystal or as a glass. Cristobalite (left), a high-temperature form of quartz (SiO_2), is similar to diamond in that it has a face-centered-cubic structure. Silicon atoms (gray) occupy the sites filled by carbon atoms in the diamond lattice. In addition an oxygen atom sits between every two silicon atoms. A conceivable glass structure (right) resembles a cristobalite structure that has been distorted. Also the three rings, each containing six silicon atoms, found in the cristobalite cell have been reconnected to form two rings with four silicons and one with eight.

worse than pure copper. In a really good conductor atoms all of the same kind are arranged in a perfect crystalline array.

This was completely incomprehensible before the invention of quantum mechanics. Between 1924 and 1926 Erwin Schrödinger, Werner Heisenberg and Max Born showed how to set about explaining a host of phenomena that had formerly been mysterious, and in the five years that followed the foundations were laid for the understanding of solids and of much else besides. One learned to say, when asking anything about electrons in atoms, molecules or solids: If you want to know what an electron does, forget about it and pretend there is a wave there. Calculate where the wave goes, and there you will find electrons.

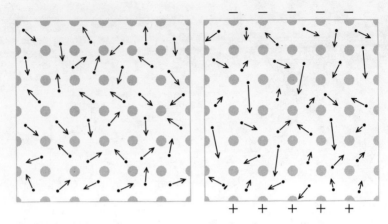

Electrical conductivity in a metal can be thought of as the movement of valence, or free electrons (black) in a preferred direction. In the absence of an external electric field (left) the movement of any one electron is offset by the movement of another in the opposite direction. Within an electric field electrons move toward the positive plate (right).

It is a well-known property of waves that they can go through a regular array of obstacles of any kind. At first this seems surprising. It is easier to grasp that waves do not go through an irregular array of obstacles. That is why the headlights of a car cannot penetrate very far into a fog; the droplets of water scatter the light out of the headlight beams. If the droplets were arranged in some regular way, as the atoms in a crystal are, this would not happen; the light would go straight through.

This property of waves, which could be proved by quite simple mathematics and had been known long before quantum mechanics, showed in principle why good conductors of electricity have to be pure, crystalline and cold. Another property of waves enabled us to understand in the early 1930's why some materials were insulators and some conductors. The explanation was first given by A. H. Wilson of the University of Cambridge. The argument did not, as one might expect, seek to tie the electrons to the atoms in nonconducting materials. For both kinds of materials Wilson started by thinking of the electrons as being free to pass through the crystal as waves. The theory went on to show, however, that in some materials there cannot be any current because there will always be just as many electrons moving one way as the other.

The solid state 27

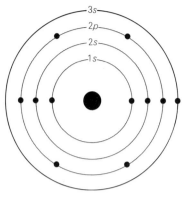

SODIUM ATOM

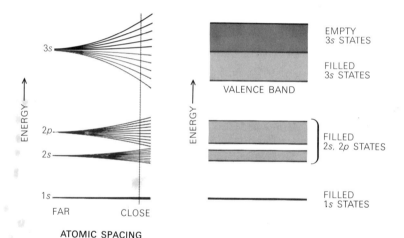

Theory of electrical conductivity involves the behavior of an atom's outermost shell of electrons, the valence electrons. Sodium, for example, has 11 electrons arranged in four shells. The three inner shells are filled, but the valence shell could hold another electron. The electrons in each shell occupy specific energy levels (bottom left). As atoms are brought close together and begin to influence one another, the electrons are forced to have slightly different energies, since only two electrons can occupy precisely the same quantum state. The vertical line indicates the spacing of atoms in a crystal of sodium. Within a crystal containing some 10^{20} or more atoms the energy levels become densely filled bands (bottom right). Since sodium has only one valence electron, only half of the energy states in the valence band are filled. As a result negligible energy is required to raise a valence electron to an empty state, where it is free to move inside the crystal and to conduct electricity.

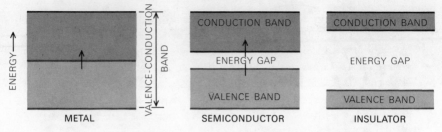

Electrical conductivity of a solid depends on the spacing and state of occupancy of the energy bands within its crystals. Many metals (left) resemble sodium in having a valence band that is only half-filled with electrons and therefore can readily act as a conduction band. Other metals have more complicated band structures but the net result is the same. In semiconductors (middle) there is a small energy gap between a filled valence band and the first permissible conduction band. It is not too difficult, however, for some of the electrons to acquire the energy needed to jump across the gap. In an insulator (right) the gap is not easily bridged.

The argument is a sophisticated one; it is based primarily on Wolfgang Pauli's exclusion principle, which says that no two electrons can ever move on exactly the same path with exactly the same speed. It is this principle that gives rise to the electron shells of atoms; there it is expressed by saying that no two electrons can have the same quantum number. Applied to metals, the exclusion principle means that electrons in a metal will have velocities lying between zero and some maximum velocity. For an insulator it happens that the limiting velocity has a value that is extremely awkward in view of the mathematical relation that exists between an electron's velocity and its wavelength. This relation is given by Louis de Broglie's formula in which wavelength equals Planck's constant divided by the mass times the velocity of the particle ($\lambda = h/mv$). The wavelength of electrons in an insulator is awkward because it has a dimension that just fits into the distance between the atoms of which the material is composed. Under these conditions the wave will become a standing wave, and such a wave describes a situation in which the movement of electrons in one direction is exactly offset by the movement of other electrons in the reverse direction.

In many crystalline materials one finds this situation in which there can be no electric current. To overcome it a considerable amount of energy is required; an electron must be hit rather hard to put it in a position where its movement is not balanced by the movement of another electron. The needed energy is gained, of course, if the material is heated to a temperature that is high enough. All solids will conduct electricity to a certain extent if they are hot; they can also be made to conduct by energetic radiation such as ultraviolet or X rays. No material can be a good insulator while it is exposed to X rays.

The theory I have just described makes use of difficult concepts of quantum mechanics, but the mathematics of it is not very complicated. It is the kind of theory a physics student learns in his final undergraduate year or first graduate one. The same cannot be said of the theoretical work that is currently being done in an effort to correct a grave omission in the theory. In the electron gas of a solid material the electrons are moving about all the time and bouncing off one another. To describe this bouncing mathemati-

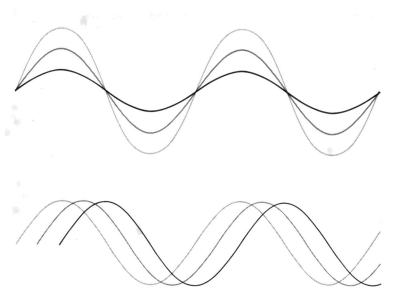

Quantum view of conductivity substitutes a wave for the motion of electrons. In an insulator (top) electron velocities correspond to a standing wave, one that does not move in any particular direction. In conductors electron velocities correspond to a running wave (bottom). The lines in the drawings show the form of the waves at successive moments.

cally is a formidable problem; it is very difficult to solve problems involving even three interacting bodies. Many-body theory is a subject of intensive research in many leading laboratories; it may be that when it is further along the metal-insulator problem will look rather different.

Meanwhile the simple model I have described has proved perfectly adequate for understanding the part of solid-state physics that is most important to technology, namely the semiconductors. These are materials that will carry an electric current but only a small one compared with a metal. Basically semiconductors should be classed as nonmetals; when they are pure and at low temperatures, they do not conduct electric current. One makes them conduct by adding electrons to them. The simplest way to do this is to dissolve in the crystal of a semiconductor traces of some chemically different material, each atom of which easily gives up an electron. Germanium, a common raw material of transistors, becomes quite a good conductor when very small quantities of phosphorus (one part in a million) are added to it. The germanium atom has four outer electrons it can easily lose; the phosphorus atom has five. It is the extra electron of phosphorus that does the trick. When it is free of the atom, it can move about quite easily, like an electron in a metal, and its motion is not offset by the motion of any other electron. Germanium that has been "doped" with phosphorus is a semiconductor of the n type, the n standing for the negative charge contributed by the additional electrons.

The technological importance of semiconductors arises mainly from the fact that the contact between a semiconductor and a metal (or between two semiconductors) acts as a rectifier. This means that the material will pass electric current much more easily in one direction than in the other. A crude form of semiconductor rectifier was used in the earliest radio receivers; it was supplanted by the vacuum tube. The replacement of the vacuum tube by the transistor represents the return of the semiconductor rectifier in refined form.

In the n type of semiconductor I have described free electrons will flow from the semiconductor to a metal but not in the reverse direction. This is hardly surprising. A cold metal does not emit electrons; it must be heated until it glows to do so in a vacuum

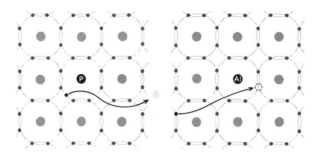

Impurity atom in the structure of germanium, a semiconductor, increases its conductivity. Each atom in a germanium crystal shares four valence electrons (dark gray) with adjacent atoms. A phosphorous atom ("P" at left) has five valence electrons. The extra electron cannot fit into the regular structure. It is in a high-energy position like the free electrons of a metal. An impurity of aluminum ("Al" at right) also enhances germanium's conductivity. Aluminum, with only three valence electrons to contribute to the germanium structure, creates a vacant site, or electron hole, into which a nearby electron can move.

tube. In the semiconductor, however, the extra electrons donated by foreign atoms can move into the metal without undue difficulty. The barrier (more exactly the change in potential energy) that keeps electrons from going from the metal to the semiconductor helps them to go the other way.

The important properties of semiconductors, then, depend on the presence in the crystal of minute quantities of some impurity. This brings me to an important point about solids. They are never quite pure and their crystal lattices are rarely quite perfect; they have what we call defects. Such defects determine many significant properties of materials, particularly the mechanical ones.

One can obtain materials with impurities present in less than one part in 10 million, but even in such materials the impurity atoms will be only 10^{-5} centimeter apart, a distance shorter than a wavelength of light. Thus quite a small speck of even a very pure material will have plenty of impurities in it. Usually the impurity atoms simply replace an atom of the surrounding crystal. This is what happens when germanium is doped with phosphorus; four of the five outer electrons of a phosphorus atom participate in bonds with germanium atoms, so that the phosphorus atom is taken into the structure of the crystal.

Other forms of impurity can impart color to a crystal, and indeed the systematic study of colored rock salt and other salts such as potassium chloride by R. W. Pohl and his colleagues at the University of Göttingen before World War II makes Pohl one of the founders of solid-state physics. It takes only a trace of blue ink to color a glass of water, and in the same way traces of potassium in an initially transparent potassium chloride crystal will make the crystal dark blue [see "The Optical Properties of Materials," by Ali Javan, page 177]. The potassium is added by heating the crystal in alkali vapor. How is the extra potassium accommodated? It turns out that some of the sites that ought to be occupied by chlorine are empty, so that adding potassium makes the crystal expand by a readily calculable amount; this has been verified by experiment. But potassium chloride is a member of the class of ionic crystals, and one cannot have such a crystal in which a large number of chlorine sites are empty. It is not an atom that normally occupies any one of the sites but an ion, in this case a chlorine atom with an extra electron stuck to it. A crystal with all those vacant sites would have an enormous electric charge. What happens is that each of the vacant sites has an electron in it. It is these electrons that give the crystal its color, by absorbing certain wavelengths of light. Vacancies with an electron in them were named by Pohl F centers, from the German word *Farbe* (color).

It is not only impurities that make a crystal deviate from perfection. In metals at high temperatures, for instance, a number of sites—perhaps one in a million—will be empty. Such sites are called vacancies. They arise because at high temperatures the atoms are vibrating, and now and then a vibration is so vigorous that a vacancy is produced at the surface; then it can move inward. Of course, vacancies will also move to the surface and disappear, and eventually a balance is set up, the number formed and the number disappearing being equal.

Vacancies move around in a crystal, much as molecules move in a gas but much more slowly. It is believed that when one metal mixes with another, which is what happens when two materials are welded together, atoms change places by jumping into vacancies, so that vacancies play an extremely important part in all the

The solid state 33

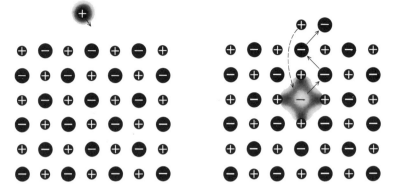

Color center arises when electrons are trapped in certain crystals. The transparent crystal of potassium chloride (left) turns blue when potassium is added to it. A potassium atom (solid arrow) attaches itself to the surface of the crystal and subsequently loses an electron (broken arrow at right). The electron trades places with a negatively charged chloride ion that has migrated outward to pair with the newly arrived potassium. The electron, which is held by adjacent positive ions, absorbs light, producing the color change.

arts of metallurgy. It would be interesting if we could see vacancies, but that is beyond the power even of present-day electron microscopes. What we can see is little clusters of vacancies. When a metal is quenched (cooled quickly), any two vacancies that accidentally meet will stick together; the vibrations of the atoms are not vigorous enough to make them move apart. Little cavities form that have curious shapes varying from one material to another, some of which are shown in the illustrations on page 34.

The technique by which these pictures are made is transmission electron microscopy, and it has turned out to be one of the most important techniques of solid-state physics. The electron microscope has the advantage that one can see objects much smaller than the wavelengths of light, which is of course not possible with the light microscope. Moreover, electrons will penetrate thin specimens of metals.

One of the great successes of this thin-film electron microscopy was the observation of another form of defect, namely the dislocation. This brings me back to the question of why a metal paper clip bends and a glass rod breaks. I remember discussing many years ago with Lawrence Bragg, the codiscoverer of X-ray crys-

Vacancy clusters can assume different shapes in different metals. In aluminum (right) *they are like disks; in gold* (below) *they are shaped like tetrahedrons. A single vacancy is too small to be visible. These electron micrographs were made by John Silcox of the University of Cambridge. The metals had been rapidly cooled. Magnification is 30,000 diameters.*

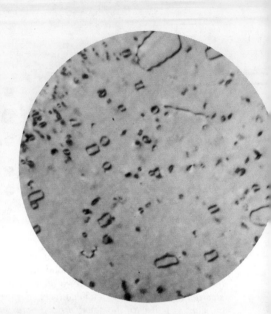

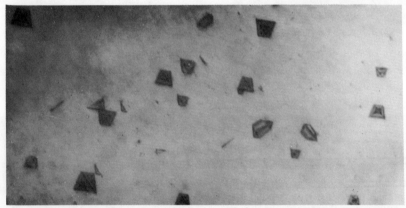

tallography, the possibility that this had something to do with the presence in metals of free electrons. It doesn't, except that metals usually have rather simple crystal structures; their atoms, having lost their outer electrons, don't form bonds with some specific number of atoms. It turns out that materials with simple crystal structures are often ductile and those with complicated ones are rarely so. Those that are amorphous are never ductile unless they are so hot that their atoms can change places quite easily. The key to the understanding of such behavior is the dislocation.

This concept was introduced in 1934 by Geoffrey I. Taylor of the University of Cambridge. The question Taylor asked himself was: When a crystalline substance is deformed, do the atoms all slip over one another together? Does the crystal pass suddenly from the undeformed state to the deformed one? For various reasons he thought that it would not, that it would instead deform through the motion of a kind of wrinkle in its structure—a dislocation in the regular array of its rows of atoms.

One of the reasons that led Taylor to put forward this hypothesis is that very pure metals are normally much softer than alloys and impure materials. The dislocation model makes it clear why impurities make a metal harder: the impurities collect in the dislocation and keep it from moving. In complicated crystal structures the dislocation itself is a complicated structure and cannot move easily. In glasses one cannot have dislocations at all.

Taylor's hypothesis explained facts known since the Bronze Age, but no one had actually seen dislocations in motion until 10 years ago. At that time some of my colleagues in the Cavendish Laboratory were examining thin metal films by transmission electron microscopy, and somewhat to their surprise they found that dislocation lines were visible. Moreover, the screen of the electron microscope showed the little lines darting forward; doubtless the metal film was buckling as the electron beam heated it up. I well remember our excitement the day one of our graduate students came into my office and said: "Prof, come and see some moving dislocations!"

Thus far I have described some of the main properties of solids that are important to the solid-state worker, particularly the electrical and mechanical properties. I have emphasized the great difference between the properties of one solid and another. There is another way of looking at these differences, namely the mechanisms by which the atoms or molecules of various materials stick together.

Perhaps the simplest of all solids are the frozen inert gases: helium, neon, argon and so on. These chemically unreactive gases do not form molecules; at any rate, if two atoms stick together, they stick so weakly that they will quickly be knocked apart by collisions with other atoms in the gas. But if you cool an inert gas, it will first liquefy and then solidify, and this shows that there is some kind of weak attraction between even the most inert atoms.

The force between such atoms is called the van der Waals attraction. It is explained by quantum mechanics, but the explanation is not simple. In general it can be said that the negatively charged cloud of electrons surrounding the positively charged atomic nucleus can slightly shift its position, so that the center of negative charge does not quite coincide with the center of positive charge. As a result of this electrical imbalance a weak force is established that can attract other atoms.

The chemically inert atoms are those in which the electrons form what is called a closed shell; such shells are found in atoms with two electrons (helium), 10 (neon), 18 (argon) and so on. Certain ions have this property too. If an electron is removed from a sodium atom, 10 electrons are left and so the remaining positively charged ion is inert like neon. By the same token, if one adds an electron to chlorine, which has 17 electrons, one gets 18 electrons and a negatively charged inert ion. An important class of solids is the ionic salts, of which sodium chloride is the best known; it is made up of inert positive and negative ions and holds together simply because the positive and negative charges attract each other.

There are solids made of atoms that are not chemically inert. Most atoms consist of an inert shell and a number of electrons in addition that can help the atom stick firmly to another atom. Carbon, silicon and germanium have four electrons outside an inert shell, and each of them can take part in forming a bond to another atom of the same kind. This kind of bonding is called "homopolar" or "covalent." The covalent bond involves the sharing of electrons between pairs of atoms.

These divisions are not hard and fast. Most of the minerals that make up the rocks of the earth's surface fall in neither one nor another. They are compounds, and the different atoms are to some extent charged; therefore they stick together partly like ionic substances such as salt. But there is a lot of electron-sharing too, and the simple classifications are not always useful.

Then there are molecular crystals. Hydrogen is the simplest example, although solid hydrogen can be obtained only at very low temperatures. The hydrogen atom has only one electron and can form a very strong bond with one other hydrogen atom, resulting in the molecule H_2. In solid or liquid hydrogen the molecules stick weakly together because only the van der Waals force holds them. Another example is water or ice, in which the H_2O molecules stick through a mechanism (the hydrogen bond) that seems to be halfway between ionic and covalent. Many organic materials are of this kind—wood and cotton, for instance. These are made up of polymer molecules, which have the form of long chains; covalent bonds link the atoms in each chain, and something like van der Waals bonds attract adjacent chains to each other [see "The Nature of Polymeric Materials," by Herman F. Mark, page 85].

Finally there are the metals. Here the outer electrons have left the atoms and can contribute to a current. The matter is discussed in the following chapter ["The Nature of Metals," by A. H. Cottrell, page 39]. Here we need only say that to obtain an adequate theoretical description of cohesion in metals is complicated; the electrons are free, they repel one another but they are attracted by the ions. Of course attraction must win—otherwise no solid metal could exist. Detailed calculations have been carried out for only a few metals, although research in this area is very active, particularly with respect to accounting for the crystal structures observed in alloys. What one might say is that it is not surprising that one can form so many alloys; metal atoms are not particular about what other metal atoms they stick to. If all the electrons come off in any case, strong cohesion exists whatever the strength of the charge is on the ion and however many electrons there are.

Reflecting on solid-state science in 1967, one perceives the following main lines of advance. In theoretical physics and in fundamental physics generally the many-body problem—the interaction of all the electrons in a metal—continues to be of great interest and abounds in unanswered questions. Allied to this subject is our recent understanding of superconductivity, the complete disappearance of electrical resistance at very low temperatures. Then the study of surfaces and of the interface between metals and semiconductors has moved into the center of the picture, partly because of its extreme importance for electronic devices and partly because new techniques for investigating them have been introduced. Finally, many solid-state physicists are looking over their shoulder at biology. Since solid-state science deals mainly with the movement of charge and energy through relatively simple solids, it ought to have a lot to say about these processes in the vastly more complicated living tissues.

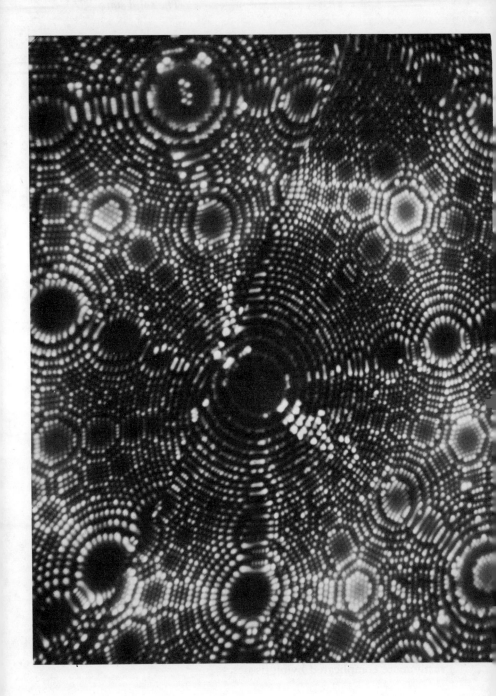

The gas of electrons that binds metal atoms together makes metals behave as they do. Their mechanical properties in particular flow from the close-packed crystal structure favored by the metallic bond.

A. H. COTTRELL

The nature of metals

Metals are opaque, lustrous and comparatively heavy. They are strong, but they can be rolled or hammered into shape and can be alloyed and welded. They are good conductors of heat and electricity. All these properties of metals flow from the metallic bond. The basis of the bond is that in a metal each atom is closely surrounded by many similar atoms, each with only a few electrons in its outer electron shell. In this situation the electron clouds overlap and the loosely held outer electrons are so completely shared as to be no longer associated with individual atoms. Leaving the metal atoms in place as ions, they form an electron gas, a pervasive glue that moves freely among the ions and binds them together.

Grain boundary between metal crystals is only a few atoms wide, as seen in a field ion micrograph. Here the tip of a tungsten needle is enlarged about 3,500,000 diameters. Each bright spot represents a tungsten atom; their pattern, which depends on the way the curved surface of the tip intersects successive crystal planes, changes abruptly at the grain boundary (bottom left to top right). The micrograph was made by J. J. Hren and R. A. Newman of the University of Florida, who used a new fiber-optical technique.

Because the electrons are free to move in an electric field, metals conduct electricity. Because free electrons absorb and then radiate back most of the light energy that falls on them, metals are opaque and lustrous. Because free electrons can transfer thermal energy, metals conduct heat effectively. The thermal, electrical and optical properties, which are responsible for many of the most advanced uses of metals, will be taken up in subsequent chapters in this book. Here I shall be primarily concerned with the mechanical properties of metals.

The metallic bond is nonspecific, which explains why different metals can be alloyed or joined one to another. It is also nondirectional, pulling equally hard in all directions. It therefore binds the metal atoms tightly, so that their cores (nuclei and inner-shell electrons) fit closely among one another. The close packing favored by the metallic bond is best realized in certain regular crystalline structures. These structures, although resistant to tension, offer less resistance to shearing forces, and thus they explain the ductility of metals. They are by definition dense, and thus they explain the comparative heaviness of metals. The mechanical properties of metals, then, derive from their crystalline structure, which is favored by the free-electron metallic bond.

As early as 1665 Robert Hooke simulated the characteristic shapes of crystals by stacking musket balls in piles, but it was 250 years before anyone could know that he had exactly modeled the crystal structure of familiar metals, with each ball representing an atom. Although the crystallinity of some substances with complex structures was recognized centuries ago, the simple crystal structures of the common metals remained doubtful until recent times. There were hints of crystallinity, such as the solidification of molten metals at a precise temperature and the bright facets sometimes seen on a fractured metal surface. Other features, however, suggested an amorphous structure. Molten metals can be cast and set in any shape, solid metals can be plastically deformed by beating them, and a polished metal surface appears to be quite featureless.

The door to the structure of metals was opened in 1864, when Henry Clifton Sorby of England developed a method for viewing metals under the microscope by reflected light instead of by light transmitted through thin specimens (the traditional biological

The nature of metals 41

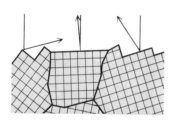

Grains in aluminum are revealed in reflected light in a photomicrograph made by G. C. Smith, S. Charter and S. Childerley of the University of Cambridge. The sample was electrolytically etched and oxidized. As shown schematically (left), some grains have crystal orientations that reflect most of the incident light and therefore appear to be bright; others reflect less light and appear darker. The magnification is 50 diameters.

and mineralogical technique). Apart from his new optical method, the key to his success lay in the removal of the polished surface layer by a careful etching treatment with a weak and nonstaining chemical reagent.

This procedure reveals an irregular honeycomb of boundaries that partition the metal into small polyhedral cells called grains, typically about .01 inch across [see the illustration above]. Some reagents etch deep grooves along the grain boundaries; others reveal the microstructure by their attack on the grains themselves. At a given angle of illumination some grains appear bright and others dark. The distribution of light and shade among the grains changes rapidly, rather like the image in a kaleidoscope, as the angle is varied, showing that the etched surface of each grain consists of small, flat, reflecting terraces all set at the same inclination to the surface of the metal. Clearly each grain, however irregular in shape, is a single crystal, and a piece of metal consists of a mass of differently oriented crystals joined together along common boundaries.

This metallographic technique of Sorby's has proved to be of lasting value for determining grain shape and size (which influence many engineering properties of metals), for identifying foreign particles and unwanted films of impurities along grain boundaries and for unraveling the microstructures that result when different metals are dissolved in one another to make alloys. Optical metallography was not able to determine the structure of the grain boundaries themselves, however. Gradually people came to the view that when a metal was cooled below its melting point, adjoining grains would crystallize as many as possible of the atoms that lay between them and so reduce the boundary to a mere interface, only about one or two atoms thick, across which the crystallographic orientation changed abruptly from that of the one grain to that of the other.

This view has been confirmed recently by the field ion microscope. In this instrument, invented by Erwin W. Muller of Pennsylvania State University, the sharp tip of a needle-shaped metal specimen is viewed end on. The tip is held in vacuum at a high positive voltage, so that lines of electrostatic force radiate from it to a fluorescent viewing screen. A trace of a gas such as helium is let into the chamber. When a helium atom touches an atom of the tip, it becomes a positive ion and then flies down the particular line of force from that metal atom to the screen. A visible image is produced by the impact of the ions on the screen. The geometry is such that an area of one atom on the tip diverges to about one square millimeter on the screen; in this sense we "see" the atomic structure of the tip. When metal crystals with grain boundaries that bisect the tip are examined by this technique, the continuation of crystal structure right up to the boundary itself, only one or two atoms thick, is clearly seen [*see illustration on page 38*].

The nature of metals 43

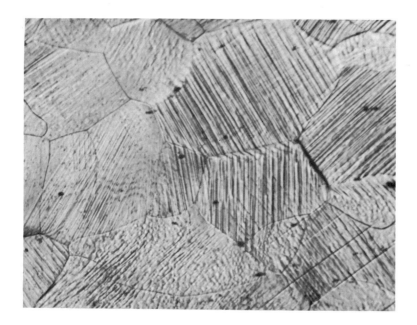

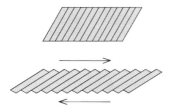

Plastic deformation reveals the slip planes in the crystals of a metal. The photomicrograph, taken by Smith, Charter, and Childerley of Cambridge, shows a deformed sample of aluminum. The parallel lines within each grain are steps formed, as with a pack of cards (left), when the metal was stressed and slip occurred on certain crystal planes in each grain. The sample has been magnified 50 diameters.

In 1900 James A. Ewing and Walter Rosenhain reported to the Royal Society of London that when a metallographic specimen was slightly deformed—for example by having its sides squeezed in a vise—its surface became crossed with fine lines. The lines generally ran straight and parallel across each grain in well-defined directions that were different in different grains [*see the illustration above*].

Careful study showed that these lines were the traces of steps formed where neighboring thin sections of the crystal had slid past one another like sliding cards in a deck. Later investigations, particularly studies of large single crystals of metals, showed that sliding occurred in certain planes of atoms in the crystals and along certain crystal axes in these planes. The mechanism by which a metal is plastically deformed was thus shown to be a new type of flow, vastly different from the flow of liquids and gases. It is a flow that depends on the perfectly repetitive structure of the crystal, which allows the atoms in one face of a slip plane to shear away from their original neighbors in the other face, to slide in an organized way along this face, carrying their own half of the crystal along with them, and finally to join up again with a new set of neighbors as perfectly as before, thereby restoring the original properties and internal structure of the crystal.

If plastic slip is a consequence of the regular structure of a crystal, why does it occur in metal crystals rather than in nonmetallic crystals such as diamond and sapphire, which generally prefer to break instead? Why, that is to say, are metals so ductile and most nonmetals so brittle? To understand this we shall have to go more deeply into the nature of metals.

Three kinds of crystal pattern are common in metals. In the body-centered-cubic structure an extra atom is packed into the center of a simple cubic cell. The alkali metals, such as sodium, and also iron at room temperature, chromium, tungsten and molybdenum have this structure. In the face-centered-cubic structure there are atoms at the center of each face of a simple cubic cell. Iron at high temperatures, copper, silver, gold, aluminum, nickel and lead have the face-centered-cubic structure. In the hexagonal close-packed structure three extra atoms are located in alternate interstices of a simple hexagonal cell. Among the hexagonal close-packed metals are zinc, magnesium, cobalt and titanium.

In both the face-centered-cubic and the hexagonal close-packed structures the atoms are packed as close as possible. Both can be built up from close-packed planes laid one on top of the other [*illustration on facing page*]. Each trio of neighboring spheres in a close-packed layer provides one hollow in which a sphere of the next layer above can rest. There are two different sets of such hollows, each providing all the sites for a close-packed layer. If the sites of the first layer are labeled *A* and those of the two sets of hollows are labeled *B* and *C*, the face-centered-cubic structure is

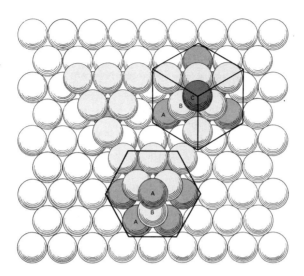

Close packing of atoms is characteristic of most metals. The two close-packed structures are modeled here by layers of close-packed spheres. On a first layer of spheres at sites A, *a second layer* (lightest gray triangles) *is placed at sites* B. *There are two sets of sites for a third layer: either at* A *(above the first layer), or at* C. *If the layers are stacked* ABCABC, *face-centered cubes are formed* (top right). *If the layers are stacked* ABABAB (bottom), *hexagonal close-packed structures are formed (these would have seven atoms in the third layer, only one of which is illustrated).*

formed when the layers are stacked in the sequence *ABCABC* and the hexagonal close-packed structure when the stacking is *ABABAB*.

In metal crystals slip occurs in directions in which the atoms are most closely packed. It is easy to see why this should be so: Close-packed rows necessarily lie farther from each other than other rows do, so they can slip past each other with less interference. Moreover, as we shall see, atoms sliding along close-packed rows pass through stable positions most frequently. Because of their high symmetry, cubic crystal structures have close-packed rows running in many directions and can therefore slip in many directions. This has an important effect on ductility. By slipping simultaneously in several directions a crystal can change into any shape that has the same volume. It can adjust its shape exactly to fit its neighboring grains without having to open up holes or cracks at the grain boundaries, and so the entire mass of grains can be reshaped arbitrarily without breaking up. The hexagonal structure, because of its lower symmetry, does not have this property to the same extent, and so the hexagonal close-packed metals tend to be more brittle and less easy to work mechanically.

The plastic properties of metal crystals thus depend largely on the simplicity of these crystal structures, which provide close-packed directions and planes suitable for slip. Why are such structures formed? To answer this question we must turn to the electronic structure of metals.

In 1900 the German physicist Paul Karl Drude proposed that a metal contains free electrons—free to move anywhere throughout the entire piece of metal—and that the flow of these electrons under an applied electric field gives rise to the high electrical conductivity of metals. The theory was later greatly improved by the application of quantum mechanics, but the basic picture remained unchanged: A gas of mobile electrons, being negatively charged, acts as a kind of mobile glue that holds the positive metal ions together by its electrostatic attraction to them. The electron gas and the spherical ions pull each other together into a compact mass, the structure and volume of which are governed largely by the geometry of close-packed spheres. When the spheres are of equal size, as they are in a pure metal, the simple crystal structures we have been discussing result. In some alloys the difference in atomic size makes for different structures and even denser packing.

Because the free electrons act as a universal glue for all the atoms, metallic crystals are largely free from the restrictions of chemical valence that are so important in most nonmetallic substances. Different metallic crystals can be bonded readily by their free electrons, so that the cohesive strength of grain boundaries in metals is very high; in fact, it is extremely difficult to break a cold metal along its grain boundaries unless the boundaries have been contaminated by impurities. This same unselectivity of the metallic bond makes it possible to join one piece of metal to another by simply bringing their clean surfaces together, as in welding and soldering. On an atomic scale it allows metal atoms of different kinds to intermingle in various distributions and proportions on a common set of atomic sites in a crystal, which explains why alloys can form over wide ranges of composition.

How does the free-electron structure determine the strength and ductility of metals? We have already seen one effect: It produces in the common metals (and also in many alloys in which the atoms are of roughly equal size) simple crystal structures with close-packed directions and planes that are geometrically suitable for slip. We might also expect, since the atoms are not bonded directly together but are merely held together by the free-electron gas, that these close-packed rows of atoms could slide past each other particularly easily, without coming apart. This is so, but the argument turns out to be surprisingly subtle.

If there were no resistance to the close-packed rows' sliding along one another, the material would have no rigidity at all. A solid is a substance that has some rigidity, however, and its modulus of rigidity is a measure of the amount of shearing force required to attain a given small amount of deformation. Consider two situations involving sliding planes of atoms [*see illustration on page 48*]. In one case (*a*) the atoms in each row are more closely packed horizontally (and the two rows are therefore more widely spaced vertically) than in the other case (*b*). It is clear that a given amount of deformation (lateral displacement of the top row) requires less force (less "climb" in terms of the illustration, less distortion of the atom's electron clouds in reality) in *a* than in *b*. In other words, the modulus of rigidity is lower for shear along close-packed planes.

To know the strength of a material, however, we must know not only its rigidity but also the amount of deformation required to initiate plastic flow—to make the atoms slip. In the case of sliding planes of atoms plastic flow begins when the atoms of the upper layer reach an unstable position where there is the maximum difference between the electronic forces holding the atoms back and those pulling them forward to the next crystal sites. The amount of deformation that brings the atoms to that point is clearly less for *a* than for *b*. Thus for two reasons—the lower modulus of rigidity and the smaller deformation required to reach the point of plastic flow—slip is more likely along close-packed planes of atoms.

This argument shows why cubic metals such as aluminum and copper are particularly ductile, but it fails by orders of magnitude to account accurately for the shear strength of metals. Calculations of the ideal shear strength, based on this line of reasoning,

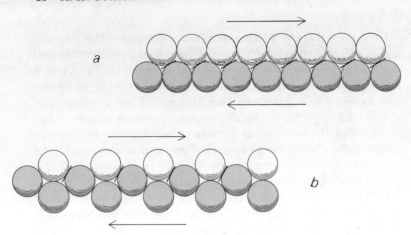

Plastic flow occurs when planes of atoms slip past one another. Close-packed planes do this more easily (a) than planes aligned in another direction (b). The atoms in a row are closer and the rows are farther apart in a than they are in b, and so less force is required for a given horizontal displacement, as suggested by the pitch of the bars. Moreover, less displacement is required to move atoms into unstable positions, from which they will be pulled forward into stable ones, when these stable positions are closer together (a).

long ago indicated that a metal should deform elastically some 3 to 10 percent before beginning to flow. In practice, however, a crystal of pure metal usually flows at deformations as low as .01 percent. This thousandfold discrepancy between ideal and real strengths is of great practical as well as scientific interest.

When the discrepancy was first discovered, it was thought that the theory of ideal strength might be wrong. It is not wrong; metal crystals grown in the form of thin "whiskers" do in fact have strengths near the ideal value. The softness of large crystals of pure metals is not due to any change in the laws of atomic forces but to the presence of dislocations in the crystals: irregularities of crystal structure that allow atomic planes to slip much more easily than they would in a perfect crystal.

Sir Nevill Mott has likened slip to the sliding of a large, heavy rug across a floor. If you try to slide the entire rug as one piece, the resistance is too great. What you can do instead is to make a wrinkle in the rug and then slide the whole thing a bit at a time by pushing the wrinkle along, thereby enlarging the slipped region behind it at the expense of the unslipped region in front of it.

Such a process is directly analogous to plastic flow caused by a dislocation. When the part of a crystal above a slip plane slides over the part below, it does not do so all at once as in the theoretical examples we have been considering but a bit at a time. During slip there is necessarily a boundary line—the dislocation—between the slipped and the still-unslipped regions. In an "edge" dislocation the dislocation is the edge of an extra vertical plane of atoms crowded into the upper, slipping part of the crystal. The dislocation line lies at right angles to the direction of slip [*see illustration on page 50*]. When the direction of slip is parallel to the dislocation line, the result is called a "screw" dislocation. Most dislocations are actually combinations of the two and tend to take the form of loops.

Dislocations are almost inevitable in a crystal, if only because of irregularities in the process of crystal growth and the fact that every grain boundary is in effect an array of dislocations. When a crystal is subjected to a shearing force, dislocations in it are made to move along slip planes. If a crystal contained a single dislocation line and no other imperfections, under stress the line would move right of the crystal, deforming the crystal by one atomic spacing at most. In reality crystals usually contain complex networks of interconnected dislocation lines as well as other defects and impurities in the crystal lattice. When the dislocations begin to move, their ends remain tied to other parts of the network or to other defects. Because the ends are anchored, the active slip planes can never get rid of their slip dislocations. In fact, the dislocations in a plane multiply when the plane slips.

The important question with regard to the strength of a metal therefore becomes: How easy is it to move a dislocation? In the slip process the dislocations are of course pushed along by the shear stress acting on the slip planes, which comes from the applied forces—tension, compression and torsion—on the material. How much shear stress is needed to move a dislocation in a slip plane? This is really two questions, one about the natural resistance of an ideal crystal to the passage of dislocations and one about the effect of blocking the paths of dislocations with foreign particles and other obstacles in real crystals.

Let us consider the natural resistance of the crystal lattice. The atoms immediately ahead of a dislocation resist its approach, since it forces them out of their stable crystal sites. The atoms immedi-

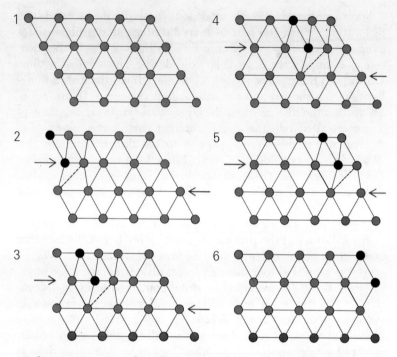

Dislocations cause metals to flow more easily than the mechanism outlined in the illustration on page 48 would predict. Dislocations move through a metal crystal in effect one line of atoms at a time. The diagrams illustrate plastic flow in a close-packed crystal (1). Under a shearing stress (arrows), a plane of atoms (black spheres) moves (2) and a bond is broken (broken black line). The extra plane (black) continues to move; bonds break and then re-form (solid black lines) as it passes (3–5). When the dislocation (region of broken bond at edge of extra plane) has traversed the crystal, the crystal is deformed by one atomic spacing (6).

ately behind the dislocation push the dislocation forward, since the farther away it gets the more completely they can fall into the new stable sites to which they have been brought by the slip process. Since the dislocation is thus pushed both forward and backward, the natural resistance of the crystal to its movement is approximately zero! This spectacular property of the crystalline state of matter holds when the region of dislocation is wide, that is, when the transition from the slipped to the unslipped areas of

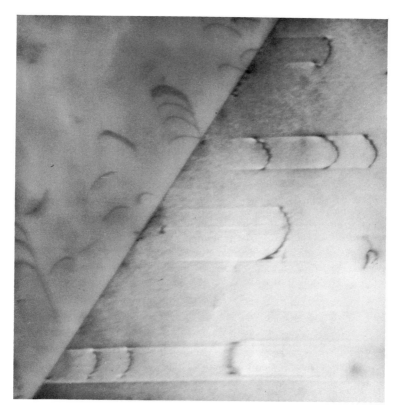

Dislocation lines in a thin film of stainless steel are enlarged 52,500 diameters in an electron micrograph made by P. B. Hirsch and M. J. Whelan of the University of Cambridge. The dislocations, which extend from the upper to the lower surface of the film, are seen as arrays of dark, curved lines that run out from a grain boundary along various slip planes.

the slip plane occurs gradually over a width of many atoms; there are then so many atoms on both sides to exert this reverse tug-of-war on the dislocation that they are always in almost perfect balance. On the other hand, when the region of dislocation is narrow, there are too few displaced atoms to maintain a good balance and an applied shear stress is needed to move the dislocation. In the limiting case of a dislocation only about one atom wide, this stress can be nearly as large as the ideal shear strength of the material.

Three dislocations are simulated by the "bubble raft" technique, in which a layer of bubbles models a layer of close-packed atoms. The dislocations, each involving an extra row of atoms, are arrayed from bottom left to top right. They form a "low-angle grain boundary"; the angle between grains can be seen by sighting along rows from the lower right.

Tangle of dislocations acts to strengthen a metal by obstructing subsequent dislocations. The electron micrograph, made by Thomas and R. L. Nolder, is of a film of nickel that has been cold-worked by rolling to strengthen it. The dislocations are concentrated in certain areas, creating a system of small dislocation-free "cells," or subgrains, in the metal.

One expects narrow dislocations in crystals such as diamond, where the highly directional covalent bonds hold the atoms at definite angles to one another and there is great resistance to shear. In agreement with this, such materials are very hard, even with dislocations. By the same token, one expects wide dislocations in metals, where the close-packed structure and free-electron bonding allow the atoms to slide over one another with relative ease. This explains the extreme softness of pure close-packed cubic metals such as copper, gold and aluminum.

The practical problem with such metals and many others, of course, is not how to make them more ductile but how to make them harder. Metallurgists do this by putting various obstacles in the way of the dislocations. Since dislocations pile up at grain boundaries, metals can to some extent be hardened by reducing the size of their grains. Alloying introduces foreign atoms that distort the crystal locally around themselves, and these local distortions offer resistance to a nearby dislocation. If the alloy atoms are gathered into clumps, their effect is enhanced, and this can be accomplished, as it is in the alloy Duralumin, by heat treatment.

In the hardening that is produced by various processes of plastic-working such as hammering or rolling, the obstacles are paradoxically the dislocations themselves. When the number of dislocations in the worked metal becomes large enough, those moving along intersecting slip planes obstruct one another's movement—an effect readily appreciated by anyone who has been held up at a road junction in dense traffic [*see lower illustration on facing page*].

We have seen that in metals dislocations in close-packed planes are intrinsically mobile and that this makes for softness and ductility. What happens, however, when a metal is attacked by a high-velocity crack? This is not an idle question. Commercial metals and alloys usually contain particles of brittle foreign substances; when such a particle suddenly splits, the metal nearby is attacked by a crack that comes out of the particle at high speed. Experience has shown that face-centered-cubic metals such as copper resist such a crack, which is blunted and changed to a wide notch by plastic deformation. The body-centered-cubic "transition" metals, such as iron, usually behave in the same way when they are warm, but they allow the crack to run through them in a brittle manner when they are cold.

Why do the face-centered-cubic metals blunt the crack? If the crack passed through the metal slowly, the dislocations already present could be moved about by the stresses accompanying the crack, and the plastic work done by their movement might drain away the crack's energy and so halt it. But what if the metal is being attacked by a high-speed crack? To understand the consistent ductility of the face-centered-cubic metals, we have to consider whether in a given specimen the atoms at the leading edge of a sharp crack are likely to pull apart, propagating the crack in a brittle manner, or to slide past one another, starting a slip process.

Anthony Kelly, Bill Tyson and I have recently looked at this problem. We calculated that the tensile forces on the atoms directly ahead of a sharp crack are about five to six times larger than the shear forces. With the approach of the crack these forces grow, but they stay in the same ratio until the failure point of the atomic bonds, either by tension or by shear, is approached. If the ratio of tensile strength to shear strength in these bonds is less than the ratio of tensile forces to shear forces, the failure should occur by cracking; if it is greater, the failure should be by slip. It is possible to estimate the approximate tensile and shear strengths of the bonds in various types of materials. When this is done, the comparison of the ratios of strengths to stresses shows quite clearly that materials such as diamond must be brittle. The issue is finely balanced for the body-centered-cubic transition metals, which agrees with the fact that these are sometimes brittle and sometimes ductile. For the face-centered-cubic metals the shear strength is so low compared with the tensile strength that these metals must always be ductile.

The low shear strength of face-centered-cubic metals follows directly from the low modulus of rigidity and the small deformation required to initiate plastic flow on their closely packed planes. These qualities, as we have seen, are implicit in the geometry of close-packed spheres that is so characteristic of the structure of metals and that is a direct consequence of the nature of the metallic bond.

Ceramics usually consist of metallic and nonmetallic atoms joined by bonds that are partly ionic and partly covalent. This gives them such properties as hardness, brittleness and resistance to heat.

JOHN J. GILMAN

The nature of ceramics

One of the first solid materials, if not the very first, that man learned to use was a ceramic, natural stone. It was suited for a variety of applications because of the characteristic properties of a ceramic: hardness, strength, imperviousness to heat, resistance to chemical attack and brittleness, which makes it comparatively easy to shape the material by chipping it. This natural ceramic provided man with tools, durable containers and even a roof over his head (in caves).

It is not surprising that man turned to creating ceramics of his own: pottery, bricks, concrete, glass. These products continue to be major industries to this day. Few people realize, however, how richly the field of ceramics has developed in this century. Ceramic materials have now been extended to a wide spectrum. They range in appearance from dull clay to lustrous ruby, and in technical service from the refractory linings of steelmaking furnaces to delicate electronic control devices. New understanding of the chemical and physical attributes that determine the nature of ceramics has expanded our view of the possibilities inherent in this class of material. Just as the understanding of structural principles and of the elements that go into the construction of a building (stone,

Oxygen and metal atoms (or semimetals such as silicon) are the basis of most ceramics. In the simplest ceramics equal numbers of oxygen (white) and metal (gray) atoms are packed together in arrangements that depend largely on the relative sizes of the ionized atoms. In beryllium oxide (top) the "coordination number" is four; each beryllium atom is surrounded by four oxygens (and each oxygen atom by four beryllium atoms, although only one is shown). In magnesium oxide (bottom) each atom has six nearest neighbors. The ceramic bond is primarily ionic: each metal atom gives up two electrons to each oxygen atom.

mortar, steel, wood, glass and so on) gives an architect wide scope for the design of structures to suit particular purposes, so the ceramist's understanding of how atoms and crystals can be put together to form the architecture of a ceramic now provides a wide range of opportunities for obtaining specific properties. The architectural analogy is a pertinent one, because the properties of ceramics depend crucially on their microscopic structure.

What is a ceramic? Essentially it is defined as a combination of one or more metals with a nonmetallic element, usually oxygen. The comparatively large oxygen atoms serve as a matrix, with the small metal atoms (or semimetal atoms such as silicon) tucked into the spaces between the oxygens [see illustration on this page]. A cardinal characteristic of the construction of ceramic crystals is that the atoms are linked by bonds that are primarily ionic but also to a significant extent covalent. These firm bonds are primarily responsible for the stability and strength of ceramic materials. In the combination of oxygen atoms with metal atoms the ionic bonds are particularly strong, because each oxygen, with two electronic vacancies in its outer shell, borrows two electrons from its metal neighbors; thus both kinds of atom become strongly ionized—one negatively, the other positively—and are held together by a correspondingly strong electrostatic attraction. In this sense ceramic compounds resemble salts, which are characterized, however, by more purely ionic bonds. The bonds are stronger in ceramics, which therefore dissolve in water only under high pressure.

As highly oxidized compounds the ceramics are strongly resistant to attack by nearly all chemicals. This accounts for many of their uses—in plumbing, household utensils and so forth. They are indispensable where a structural material must withstand very high temperatures, at which the ravages of oxygen will destroy the strength of a metal. Indeed, the making of steel depends on the use of ceramic bricks to line the inside of the container.

Since the art of ceramics-making has a long history, it is of interest to review its evolution briefly before we examine the present state of the art and its future prospects. The first man-made ceramics probably were simple mud pots, sometimes reinforced with straw, that were dried and baked in the sun. Gradually the ancients learned to improve the product by selecting suitable clays and baking their pottery at higher temperatures, at first in open fires, later in crude kilns. These early earthenware pots must have been disappointingly porous, and it was a large step forward when some clever artisan in Egypt learned to seal the surface by firing a glassy material onto it. With this step, and the addition of natural stains to color the glaze, ceramics became an aesthetic art as well as a useful one. The art grew in beauty and utility as people gained increasing control of it by grinding rocks into powder to produce more uniform clays and built kilns that provided higher firing temperatures, making possible the production of a wider variety of ceramic materials.

Scientific knowledge of ceramics began with chemical analysis of the components that make up these materials. Chemists found that the principal elements in natural clays were oxygen, silicon and aluminum, forming the compounds known as aluminosilicates. The precursor of common clay is feldspar, a mineral found in rocks such as granite. It usually contains potassium and has the chemical formula $K_2O \cdot Al_2O_3 \cdot 6SiO_2$. Rain and carbon dioxide in the atmosphere convert the feldspar crystals into potassium carbonate (which dissolves and is washed away) and silica and aluminum oxide. These combine with water and usually form kaolinite, whose basic formula is $Al_2O_3 \cdot 2SiO_2 \cdot 2H_2O$. This material is often mixed in the soil with small amounts of iron oxide, which gives it the reddish color of common clay (pure kaolinite is white and is the basis of fine china).

The crystals of kaolinite are very small—thin platelets only about 5,000 angstrom units (.00005 centimeter) wide and about 300 angstroms thick—and have an irregularly hexagonal shape. When wet, they cling together like wet sheets of paper, and with water acting as a lubricant the crystals readily slide over one another; this accounts for clay's plasticity. As the clay begins to dry, the crystal platelets tend to become interlocked, and the clay object becomes rigid enough to be removed in one piece from the potter's wheel or mold. The firing of the clay in a kiln evaporates most of the remaining water, and during the process some of the silica (SiO_2) combines with impurities to form a liquid glass that glues the crystal platelets together. Thus a clay ceramic becomes a kind of microscopic concrete, consisting of a "gravel" of aluminum silicate particles held together by a glassy cement.

Not all ceramics are crystalline. Glass, of course, has a noncrystalline structure, and that distinctive form of ceramic is discussed in the next chapter. Nor are all ceramics compounds of oxygen; certain other substances, among them graphite, diamond, silicon carbide (Carborundum), tungsten carbide and uranium carbide, also have ceramic properties, such as high mechanical strength and resistance to heat and chemicals. For illustration of the basic nature of ceramics, however, we shall confine ourselves here to the crystalline, oxygen-containing compounds, which are the typical and by far the most common ceramic chemicals.

The secret of these ceramics' properties lies as much in the internal structure of their crystals as in their chemical composition. A ruby and a rough brick may both be made of the same substance—aluminum oxide—yet how different they look and how differently they behave! The difference is that a ruby consists of one large crystal with its atoms arranged in a periodic pattern, whereas the brick consists of a collection of many crystals cemented together.

A giant step forward in ceramic science came when crystal structures were analyzed by means of X-ray diffraction. The findings led to detailed calculations by the physicist Max Born of

Silicate unit is a primary building block of many ceramics. It consists of a silicon atom surrounded by four oxygen atoms. This is the same tetrahedral arrangement as in beryllium oxide (see illustration on page 56). Since each of the silicon atoms has four valence electrons to give up, each of the surrounding oxygens gets one, leaving its outer shell one electron short. It can get that electron from another silicon atom by linking two groups (left). In this way a chain of silicate groups can be built up (right).

electrostatic forces in crystals. He showed that the atoms in crystals, in particular those in oxides, are bound together in accordance with Coulomb's law: the electrostatic attraction, or binding force, between two ions of opposite charge increases in inverse proportion to the square of the distance between them. Consequently the more closely the atoms are packed, the greater will be the crystal's elastic strength and resistance to alteration by heat or chemical assault.

It follows, then, that in the construction of a ceramic material much depends on how the atoms are stacked. If the growth of a crystal is closely controlled, layers of tightly packed oxygen atoms form with small metal ions in the crevices between the oxygen atoms; then another closely packed oxygen layer forms above the first, and the process continues. In this way a highly stable structure can be built. The simplest structural forms are produced during the growth of crystals consisting of equal numbers of oxygen and metal atoms, such as beryllium oxide, magnesium oxide or zinc oxide. When the metal atoms are outnumbered by oxygen atoms, as in aluminum oxide (Al_2O_3), the holes between the oxygens are not all filled; in this particular case the aluminum atoms occupy only two-thirds of the available spaces.

The most interesting and important structures are generated by silica (SiO_2). The silicon atom, like carbon, has four valence electrons, and it forms a tetrahedral grouping with oxygen: four oxygen atoms surrounding each silicon atom. These groups can link together in various ways [*see illustration above*].

Attached only end to end (by way of one of the oxygen atoms), they form a fiberlike chain, such as appears in asbestos. Built up in sheets, they produce layered minerals such as talc or mica. They can also be linked to produce a three-dimensional network — the quartz crystal. The versatility of the silica tetrahedrons in forming bonds with one another and with other groups explains how silica serves as the glue that cements the clay particles in bricks and earthenware and bonds the glaze to porcelain.

When the silica tetrahedrons link up in a three-dimensional structure, the structure may have holes large enough to accommodate atoms of various kinds. By filling the holes with sodium and sulfur atoms one can produce the beautiful blue stone known as lapis lazuli (which in powdered form is the pigment ultramarine). The holes in a silica layer can also serve to bind it to an entire layer of a quite different but complementary structure. This is how kaolinite clay is organized. In kaolinite the water is dissociated into hydrogen and hydroxyl ions. The hydroxyls, bound together by aluminum ions, form a rather loosely packed layer, and a set of its nearly spherical ions mesh into the pattern of holes in a silica layer, thus putting the two layers in intimate contact. The aluminum ions nestle into cavities between these layers; as a result each aluminum ion is surrounded by six closely adjacent oxygen or hydroxyl ions [see facing illustrations]. It would be difficult to design a more sophisticated atomic architecture.

If the ceramic structures were perfectly organized and uniform, ceramic materials would be a great deal stronger than they are. They are greatly weakened by the irregularities that occur at crystal boundaries — much more so than metals. These irregularities are of three kinds. Local separations, or voids, may occur between the crystals, with the result that atoms can wander through the spaces, gases can permeate the material and the crystals can slide past one another. Because of this ceramics tend to flow at very high temperatures like wet sand. A second cause of weakness at a boundary is that if one crystal is out of line or twisted with respect to its neighbor, the bonds between them may be stretched or otherwise disrupted. The third source of trouble is that ions with the same charge (positive or negative) may be brought close together; the consequent electrostatic repulsion produces strain in that region of the material and may generate cracks.

The nature of ceramics 61

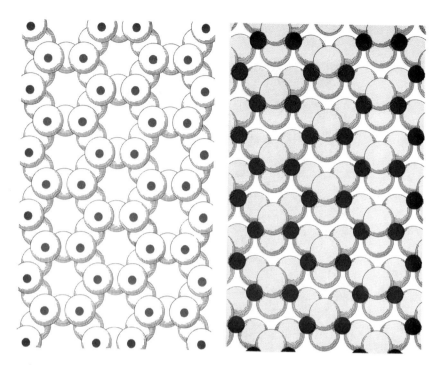

Silicate sheet formed by linked chains is the basis of many minerals (left). *Each silicon atom* (gray) *is surrounded by four oxygens; each tetrahedron shares three of its oxygens with three other tetrahedrons. Notice the hexagonal pattern of "holes" in the sheet. An aluminate sheet* (right) *consists of aluminum ions* (dark gray) *and hydroxyl* (OH) *ions* (light gray). *The top layer of hydroxyls has a hexagonal pattern. If the two sheets are superposed (as if they were facing pages of a book), they mesh, forming kaolinite.*

Kaolinite sheet, here seen from the side, is a "laminate" of the silicate and aluminate sheets shown above. In kaolinite each of the aluminum ions becomes surrounded by six close neighbors: oxygen ions and hydroxyl ions.

62   John J. Gilman

Quartz is a three-dimensional structure of silicate units. In this electron micrograph quartz crystals are enlarged approximately 8,000 diameters. Since such crystals are opaque to the electron beam, a "replica" technique is used: the specimen is first shadowed by evaporating platinum vapor onto it at a shallow angle. Then a layer of carbon is deposited, forming a replica of the surface. Finally the clay is dissolved away in acid and the replica is micrographed. Micrograph by Joseph J. Comer of the Sperry Rand Research Center.

Because boundaries between the crystal particles are a highly characteristic feature of ceramic materials, ceramists have given careful study to their properties, and much current research is directed toward eliminating the internal boundaries or minimizing their effects.

To complete our summary of the basic structural features of a ceramic we need to consider another type of imperfection. As other chapters in this book point out, all solid materials are marked by structural imperfections that affect their dynamical properties, that is, the flow of heat, electricity, mass, magnetism and so

Kaolinite crystals are enlarged 19,000 diameters in this electron micrograph made by Joseph J. Comer of the Sperry Rand Research Center. The technique was the same as used in making the micrograph of quartz on the facing page, but in this case the image was printed as a negative.

forth within the solid. An important class of these imperfections is atomic vacancies, or holes, in the crystal structure. In the case of a ceramic such holes are essential for the consolidation of the material in the hardened form, as the particles must change their shape (which involves a movement of atoms) in order to become tightly packed in the firing process.

The presence of vacancies in ceramic crystals can be demonstrated experimentally by placing crystals in an electric field. Because they have no free electrons most ceramics are insulators— highly resistant to the flow of electricity. They can, however, conduct electricity to some extent by movement of their ions, and this occurs when a hot ceramic is placed between two electrodes connected to a battery. Metal ions in the crystal (which are usually more mobile than the larger oxygen ions) may jump to the surface near the negative electrode; the jumps are made possible by the normal thermal agitation of atoms in a hot crystal. Other metal ions can then jump into the vacant sites left behind. Thus the holes move toward the positive electrode and the crystal gradually becomes polarized, with a preponderance of metal (positive) ions near the negative electrode and a surplus of oxygen (negative) ions left near the positive electrode.

If the battery is then disconnected, the ions will gradually diffuse back to a balanced distribution in the crystal. On the other hand, if the battery is left on and its voltage is high enough, some of the metal will become completely separated from the oxygen and will be concentrated near the negative electrode. Note that the behavior of the metal ions in the ceramic is like that of the positive ions of a salt dissolved in water, which also travel to a charged negative electrode. There is an important difference, however. In the salt solution the negative ions migrate just as readily as the positive ions, whereas in most ceramic crystals the metal ions are much more mobile than the oxygens. This means that a ceramic is differentially permeable to one species and can therefore be used to separate charged particles. This property of ceramics promises to make them useful in fuel-cell devices.

The basic knowledge that has been gained about the microstructure and the physical and chemical behavior of ceramic materials not only has provided a set of principles for their construction but also has led to the development of new techniques for making them. By careful attention to the control of chemical composition, particle sizes, uniformity, purity and the arrangement and packing of atoms it has become possible to synthesize high-performance ceramics in almost limitless variety. Electric kilns have been developed to provide precision in the firing process. Several new methods have been devised to synthesize ceramics in special forms. Among these are the use of very high pressure to squeeze and consolidate fine particles prior to firing; the decomposition of ceramic chemicals at high temperature to deposit a coating or shell on a background material; the use of a plasma gun that liquefies fine particles at high temperature and sprays a paint-like ceramic coating on objects; the precipitation of ceramics from aqueous solutions at high pressure and temperature (called hydrosynthesis); the slow solidification of liquids to form individual ceramic crystals with ideal properties.

To illustrate the potential versatility of ceramic materials one need only enlarge a little on the case of aluminum oxide, to which I have already referred. I have mentioned gem rubies, which are high-quality crystals of aluminum oxide containing a small amount of chromium oxide, which colors them red. The same basic substance (Al_2O_3), with titanium atoms added, yields blue sapphires. Individual crystals of aluminum oxide have great utility as watch bearings, phonograph needles, pressure-resistant windows and other products. In transparent form multicrystalline alumina ceramics are employed in place of glass as envelopes for lamps and electronic devices. A thin film of aluminum oxide is often formed electrochemically on aluminum metal to protect it. As a loose powder (corundum) it is a hard, heat-resistant abrasive. In the form of fine particles bonded together by silica glass, aluminum oxide is the basis of spark-plug insulators, refractory bricks and crucibles for molten metals. Consolidations of aluminum oxide powder are also employed as electric insulators and windows or radomes transparent to microwaves.

Without attempting to survey the entire voluminous catalogue of ceramic materials now available, I shall indicate their range by mentioning examples of various groups that can be classed according to their salient properties.

One large group exemplifies properties in the field of optics. In that area, as we have already seen, ceramic techniques are notably useful for the coloring of materials, including paints and plastics. A ceramic pigment is exceptionally durable because it is completely oxidized and hence not subject to chemical attack or deterioration. Various metallic oxides can serve as the pigments, each yielding a specific color because of the metal's selective absorption of light. (The cobalt ion, for example, absorbs red light, thereby giving a bluish tint to glass.) Coloration can be introduced into a ceramic in various ways. A salt containing the chosen metal can be dissolved in molten glass and applied to ceramic objects as a colored glaze. Alternatively, the color can be incorporated in the

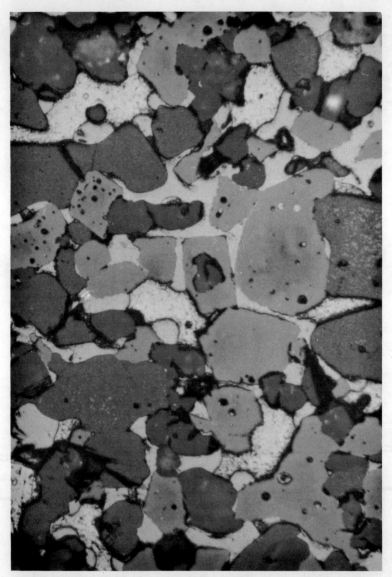

Structure of a ceramic brick used in copper production is magnified 240 diameters in a photomicrograph made by H. A. Freeman of Harbison-Walker Refractories Company. Constituents (in order of increasingly light gray) are magnesium oxide, spinel, copper oxide and copper.

form of very small colored crystals embedded in a glasslike matrix. These crystals are often spinels (normally compounds of magnesium, aluminum and oxygen) in which metal atoms producing the desired color are substituted for the magnesium or the aluminum. The color sometimes depends on the concentration: a small amount of chromium, in place of aluminum, colors spinel red, but a larger concentration of chromium will color it green. Still another way to color is to precipitate colloidal particles of a metal in a glass. The particles selectively scatter light to yield rich colors. For example, gold particles about 400 angstroms in size provide the color in the ruby glass of stained-glass windows.

There are many other optical applications of ceramics; three examples will illustrate their diversity. Calcium tungstate (found in the mineral scheelite) is fluorescent and has long been used to make blue screens for oscilloscopes; in the form of individual crystals it has now been found to be, like ruby, an excellent material for lasers. In the field of decoration, thin films of metal oxides are coated on glazes to produce an iridescent effect, and objects so coated are called lusterwares. And not the least interesting of the optical ceramics is a new material (called Yttralox) that is as transparent as window glass and can resist far higher temperatures. It is the product of a new technique based on the microscopic study of ceramic structure. Ceramics are generally opaque because of the presence of tiny pores within them that scatter light, and the new technique serves to eliminate these pores during the firing stage.

Methods have also been developed to improve the refractory quality of ceramics, one for which they have traditionally been famous. One of these advances is the fabrication of beryllium oxide (a highly toxic material), which has an exceedingly high melting point and serves as a useful material in some nuclear reactors. It has also become possible to reduce the brittleness of ceramics—a traditional shortcoming. By prestressing them, particularly at the surface, ceramic materials can be made resistant to breakage. Among the products of this improvement are "unbreakable" dishes and stronger structural materials.

In the field of electronics and magnetism ceramics are launched on a new career. They have long served, of course, as electrical insulators. Now they play more active roles based on certain unusual electrical and magnetic properties. One of these is the piezoelectricity of quartz crystals: the electric fields produced when the crystals are compressed. This property makes it possible to excite and detect the mechanical oscillations of the crystals electrically. The crystals are therefore used for precise control of the frequencies of oscillators and for the generation of high-frequency sound waves. The precise electrical measurement of the frequencies of mechanical oscillations in quartz crystals is the basis of sensitive thermometers and weighing balances. Crystals related to magnetite (iron oxide) are useful as components in high-frequency electronic elements and memory devices in computers.

Ceramics may become a much more significant factor in the electronics field than they are at present. They have many properties in common with well-known semiconductors such as germanium and silicon. Further technical development may reduce the distinction between them. Some engineers already regard semiconductors and ceramics as provinces of the same industry.

The proliferation of ceramic products is dramatically reflected in the current production figures for the industry. If we exclude glass and cements, which are major industries in themselves, the total production of the new ceramic products in money value already outweighs the manufactures of the more traditional materials—essentially building materials and pottery wares.

This trend is certain to continue. The advances in theoretical understanding of the solid state and in engineering manipulation of materials are giving rise to an ever increasing diversity of new ceramics. The movement in this field is toward increasing simplicity and more rigorous control of the materials: toward purification of the component elements, simplification of the internal structure and refinement of the production techniques. At the same time research is going forward on combining the refined building blocks into composites with desirable properties. What began as the synthesis of rocks and gems by men has developed far beyond the achievements of nature. The elements can now be fabricated into superior materials that have the heat resistance to transport us to the planets, the strength to explore ocean depths and a response to electric fields that can condense the world through the medium of optical communications.

The geometry of glass structure is the geometry of disorder on the way to order. The art of the glassmaker can be explained in terms of thermodynamics, chemical bonding and molecular architecture.

R. J. CHARLES

The nature of glasses

The study of solids is largely a matter of geometry. This is so because the geometric arrangement of the building blocks of solids is simply an expression of the binding forces between the assembled units. It is these forces that determine physical and chemical properties. The building blocks of a solid can be considered as arrays of atoms, clusters of molecules or simply single atoms or molecules. The properties of the solid depend on the placement or arrangement of these units over distances that vary from atomic dimensions to perhaps several centimeters. Many precise techniques are available for studying the atomic arrangement of matter in crystalline solids. These techniques generally depend on the scattering, diffraction or refraction of energetic radiation. In order for the resulting pattern of radiation to convey information about atomic positions the structure must be periodic, or regular, over distances that are large compared with the wavelength of the radiation. We are concerned here with glasses, a class of solids that do not crystallize when cooled from a melt and thus do not exhibit long-range periodicity of atomic structure. They yield their atomic configurations only slowly to the usual methods of structure analysis.

The structure of a glass is often inferred from the analysis of some crystalline modification of the material that forms it. Information on the disordered structure of glass itself, however, can be obtained by thermodynamic measurements. Such information cannot be specific since it is expressed in terms of a few simple variables that are themselves averages. The approach is nonetheless a useful one because in a system as complex as a glass average characteristics should have unusual significance.

Since a glass is distinguished from other solids by its lack of crystallinity, it is pertinent to examine how crystallization normally takes place when a liquid is cooled. Initially let us consider a pure liquid, that is, a liquid whose composition will not be changed by the freezing process. As a further restriction we shall require that the cooling be sufficiently slow for no changes in the liquid to occur if the cooling is halted for some period of time.

The main effect observed in cooling is that the change from a liquid to a crystalline solid is usually abrupt and occurs at a particular temperature. One might reasonably ask: Why that temperature and no other? Close observation will reveal not only that heat is passed continuously from the liquid to its surroundings during cooling but also that at the freezing temperature the surroundings receive a sudden burst of heat from the liquid.

This burst of heat is a consequence of the process by which atoms or molecules, initially in some state of chaos in the liquid, form the geometric arrays found in the crystal. One can appreciate that if there are attractive forces between particles in an assemblage, then energy, as heat or work, may be released if the particles relax into a closer and less random association with one another. Even though one can see no particularly marked change in the liquid during the initial stages of cooling, it must be assumed that the steady contraction of the liquid and the release of heat indicate that the molecules forming the liquid are engaged in a continuous ordering process. This ordering consists of many factors, such as the formation of new molecular aggregates and the development of a more geometric organization of atoms or aggregates of them, together with a reduction in the amount of random oscillation of atoms around mean positions.

At this point one might ask: Is there a simple means of measuring the amount or degree of ordering of atoms or molecules in

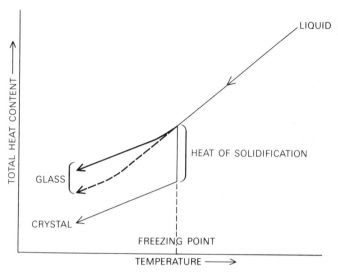

Solidification of liquid can lead to either a crystal or a glass. When a liquid solidifies into a crystal, it gives up a burst of heat at some particular temperature, which identifies its freezing point. This burst, or heat of solidification, coincides with the final ordering of atoms or molecules into a crystalline array. Liquids that cool to the rigid state without crystallizing are called glasses. Glasses cooled slowly (heavy broken curve) more closely approach crystalline state than glasses cooled rapidly (heavy solid curve).

a process such as freezing? Certainly the amount of heat given off is related to the amount of ordering that occurs on crystal formation, and one might expect a simple proportionality between the two. There is, however, an important difference between the heat transferred at a high temperature and the heat transferred at a low temperature: the first is a much rarer and more useful commodity. Assuming that the geometric arrays produced by freezing are fairly independent of temperature, a suitable measure of the amount of geometric ordering that occurs on crystallization can be provided by the ratio of the quantity of heat transferred to the temperature at which transfer takes place. This ratio, which can be put on a firm thermodynamic basis, is the measure we seek. Physicists will recognize it as entropy.

In applying this measure one can immediately see that the heat received by the surroundings on the freezing of a liquid must produce a degree of structural change in the surroundings equal and opposite to the change that occurred in the liquid. Thus the net ordering (or disordering) that occurred in the total system composed of the liquid and its surroundings suffered no change during freezing. It is a consequence of the second law of thermodynamics that this is so for a system that changes yet has no capability for doing useful work during the change. If a net change in order had taken place in the freezing process, then we could say with confidence that the change could only have been in the direction of decreased order, because processes that might give rise to a net increase in the state of order of the universe are, as far as we know, impossible. These are the thermodynamic requirements that establish the temperature at which a particular liquid can freeze; at no other higher temperature can the geometric ordering required for crystal formation occur and leave the order of the universe undisturbed.

Let us now consider how the freezing process is affected when foreign atoms or molecules are added to a liquid. Let us suppose these foreign atoms are relatively inert in that they interact with the atoms of the original liquid in much the same way that they do with one another. When the solution is cooled to produce the same crystals as before, the atoms that form the crystal not only must become organized in the same geometric fashion but also must undergo an additional degree of ordering by first being unmixed from the foreign atoms. Although the total ordering in the solution that is forming crystals is thus greater than it is in the pure liquid, the amount of heat passed to the surroundings at the freezing temperature (per unit weight of crystal formed) remains the same.

The only way for the ordering in the surroundings to be equal and opposite would be for crystals to form at a lower temperature than the one needed to freeze the pure liquid and for the liberated heat to pass to the surroundings at the lower temperature. This follows because a lower transfer temperature means a smaller number in the denominator of the fraction, or ratio, mentioned earlier that indicates how much ordering can occur on crystallization. Since the amount of heat transferred (the numerator) is unchanged, reducing the size of the denominator yields a larger value for the amount of ordering (or equivalent disordering) that can be produced. This argument illustrates the important phenomenon of freezing-point depression, which governs the freezing of seawater and the melting of road ice by salting and aids greatly, as we shall see, in the formation of complex glasses.

What I have just described are the conditions necessary for a liquid to be converted into a crystalline solid. Cooling to or below the freezing point is not, however, a *sufficient* condition for crystallization. I have simply dealt with the characteristics of assemblages of atoms before and after solidification. The mechanism for bringing about such a transformation is another matter.

Before a liquid can crystallize it must have in it a seed—a tiny crystal. A seed often consists of atoms or groups of atoms that have become attached to foreign particles or to irregularities on the surface of the container holding the liquid. Under certain circumstances, however, small numbers of atoms will spontaneously aggregate to form a tiny crystal nucleus on which other atoms can then deposit. In much the same fashion that energy is stored in the surface of a bubble, energy is stored by those atoms on crystal surfaces that find themselves in an environment that is partly liquid and partly solid. If the growing nucleus is small, the energy (per unit weight of nuclei) may be appreciable, because the exposed surface of the nucleus is so large with respect to its mass. This energy, which is stored on nuclei surfaces and is in excess of the heat energy that must be evolved for solidification, can be obtained only if the temperature of the liquid is lowered somewhat below the thermodynamic freezing temperature. In other words, a certain amount of supercooling is needed to form the nuclei that trigger the process of crystal growth.

Now let us turn our attention to time factors, which often play a crucial role in determining whether or not such nuclei, even under supercooled conditions, can form and grow. So far the discussion has been limited to a situation in which the freezing process takes place so slowly that the molecular configurations attained are determined only by temperature and not by the time elapsed in reaching any given temperature. Because molecules must slide past one another to change their configuration, it is evident that the time required for molecules to assume new relations depends on the ease or difficulty of the sliding. For most common liquids, which are composed of individual atoms or molecules that are more or less spherical, sliding proceeds easily and the attractive forces developed among atoms or molecules during cooling are more than adequate to move them into place at normal rates of cooling. In such liquids crystal nuclei form and grow easily with supercooling of only a few degrees centigrade.

Certain liquids, however, become particularly viscous near the freezing point, and the formation and growth of crystal nuclei may be prevented even when the cooling rate is slow. These liquids, unable to form or grow nuclei, follow the supercooled route to the glass state. In addition, because their viscosity increases as the temperature falls, their molecular configuration lags further and further behind the temperature. As a result the molecular arrangement at any instant will correspond to equilibrium for a temperature much higher than the actual temperature.

If cooling is continued until the glass becomes rigid, random structures characteristic of liquids at much higher temperatures will be frozen into the structure of the glass. Moreover, the faster the rate of cooling, the higher the temperature to which these frozen-in liquid states will correspond. Since the amount of relaxation, or ordering, achieved by the molecules on cooling is strongly dependent on cooling rate, the amount of heat given off will likewise depend on cooling rate. Thus we see that glasses are rigid solids whose atomic structures, and hence properties, depend not only on composition but also on thermal history. For this reason close attention is paid in glassmaking practice to schedules of quenching (rapid cooling) and annealing (slow cooling).

From the foregoing, one can adduce two important conditions that favor glass formation. These conditions are not exclusive and may in practice require proper balancing. On the one hand the glassmaker understands that he should select complex or impure solutions in order to depress as far as possible the freezing point of any crystal that might tend to form. This procedure has been used empirically for millenniums; many ancient glass compositions correspond to regions on composition-temperature diagrams where the freezing point is at a minimum. On the other hand the frictional forces inhibiting the formation of new molecular configurations in the liquid should be as high as possible. The glassmaker can achieve this either by fast cooling or by choosing materials that exhibit inherently high viscosity.

Simple mechanical considerations suggest that a particular molecular configuration should be more conducive to high liquid viscosities than any other. This configuration is a flexible chain or, when the chains are cross-linked, a net. In order for a chain to form, individual atoms must link up with at least two other atoms. The number of linkages specifies the "coordination." For a simple chain the coordination is two: each atom is joined to one on each side. If the chain is to have side groups or if it is to be cross-linked to other chains, at least some atoms in the chain must have a coordination of three or four.

Such low coordination is favored by covalent bonds, that is, by bonds created when atoms share electrons. Atoms bonded in this way exhibit a high degree of directionality. An examination of the periodic table shows that the elements most likely to form chains by covalent bonding are in the higher-numbered columns. These are the elements whose outer electron shells are shy only a few electrons. These elements, therefore, would rather share electrons than part with them. The Group VI elements (oxygen, sulfur, selenium and tellurium) are particularly good candidates for chain formation because the addition of only two electrons makes their outer shell complete. Thus they might easily exhibit a covalent coordination of two.

X-ray-diffraction analysis has shown that when selenium and tellurium are packed in hexagonal crystals, they consist of continuous spiral chains aligned along one axis of the crystal. Crystalline sulfur and another form of selenium form chainlike rings, usually of eight atoms, that pack together in various geometric patterns. Some of these rings open into extended chains when sulfur and selenium are melted; if the melt is cooled fairly fast, these elements, as well as tellurium, become supercooled and form glasses [see illustration on page 76].

In spite of considerable study it is still debatable whether or not elemental glasses can be prepared from various near neighbors of the Group VI elements, for example boron, silicon, phosphorus, germanium and particularly carbon. Amorphous forms of most of these elements are known, but this description may simply reflect the inability of present methods to detect crystalline regions less than 20 angstrom units in size. It is well known that when carbon is combined with other elements, particularly hydrogen and oxy-

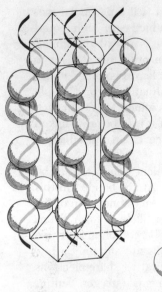

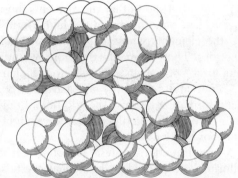

Glass-forming substances, when crystallized, often take the form of spiral chains in a hexagonal array (left) or nests of eight-member rings (right). Selenium and tellurium exhibit the spiral structure. Selenium also crystallizes in the ring structure, as does sulfur. When such ring structures are heated in a melt, the rings tend to open and link up into extended chains. If the melt is quickly cooled, the rings do not have time to re-form and a glass results. Glasses are also readily produced from spiral-chain arrays.

gen, an almost unlimited variety of chain or polymeric structures can be created. Most synthetic polymers have a glass state, and many natural organic compounds, such as alcohols, glycerol and glucose, can be supercooled to form glasses.

The physical and chemical characteristics of nitrogen, oxygen and fluorine suggest that they too may exist in the glass state, but I am not aware that this has yet been demonstrated. The well-known role of oxygen in glassmaking is to act as a glass-former by establishing stable bonds (mainly covalent) with small, multivalent ions such as those of silicon, boron, germanium, phosphorus or arsenic. The combination of oxygen with these ions yields low-coordination polyhedrons—primarily tetrahedrons or triangles—

The nature of glasses 77

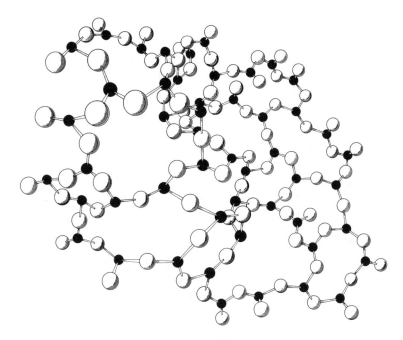

Pure oxide glass consists of a random three-dimensional network in which each oxygen atom (white) is bonded to two atoms of a metal such as boron. Here each metal atom is bonded to three oxygen atoms. However, there are other kinds of glass, such as silica glass, in which each metal atom is bonded to four oxygen atoms, producing a more complex network.

that have oxygen atoms at the corners and a multivalent ion at the center. The polyhedrons link flexibly with each other through the corner oxygen atoms. Pure oxides produce a completely crosslinked network because every oxygen atom is linked by electron-sharing to two positive ions [*see illustration above*].

Such oxide glasses, or their crystalline counterparts, are very stable and have relatively high softening temperatures. The crosslinking can be reduced and the softening temperature lowered by introducing into the melt metals such as sodium or potassium, which shed electrons easily and thus form fairly weak ionic bonds with oxygen that lack directionality. These additives, called fluxing ions, are used extensively in glass production to lower melting

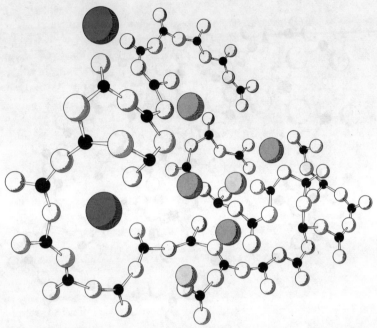

Flux-containing glass also consists of a random three-dimensional network, except that flux atoms such as sodium (gray) have reduced the amount of cross-linking. Thus some oxygen atoms are now strongly bonded to only a single atom and have weaker ties (not shown) with one or more flux atoms. As a result the melting point of the glass is reduced.

temperature, inhibit crystallization and control fluidity [*see illustration above*].

Because I have been discussing glasses in terms of atomic structure, I may have given the impression that macroscopic structure in glasses—on a scale, say, of 100 to 100,000 atomic diameters—is simply a continuation of a much finer structure and so contributes little to the properties of glass. Such is not the case.

The addition of fluxing ions to oxide or elemental glasses frequently causes the components of a melt, on cooling, to separate into two or more distinct but intermixed liquids of markedly different chemical composition. When these different liquids are frozen into a rigid glass, the way they are distributed can have a profound effect on properties such as mechanical strength, electrical conductivity, chemical resistance and optical clarity.

This oil-and-water behavior is quite remarkable because there are many cases in which a melt successfully passes a temperature at which crystals might ordinarily precipitate but does not pass a lower temperature without forming two liquids. Yet the process by which one liquid nucleates and grows out of another is very similar to the process by which a crystal nucleates and grows out of a solution. This separation phenomenon seems to be a general characteristic of chain structures and is now of recognized importance in glass technology.

Although separation mechanisms have only recently received close attention, and much remains to be learned, it is possible to describe some features of the process in general thermodynamic terms. Consider a system that consists of one beaker, A, containing chains of molecules and a second beaker, B, containing the same type of chains to which a considerable number of fluxing atoms have been added [*see illustration on page 80*]. In beaker A the entangled chains are essentially continuous and are held together only by weak forces between chains. In beaker B the chains are broken up into short segments by the fluxing ions; moreover, the electrons given up by the fluxing atoms have migrated to the ends of the segments. Thus the liquid in B is held together not only by weak forces between chains but also by the stronger ionic forces between the positive flux ions and the negatively charged chain ends.

Now imagine that a single molecule is transferred from a chain in beaker A to a chain in beaker B. This molecule, removed from a condition in which its highly directional covalent bonds were distorted to conform with entangled chains, now finds itself in a short segment of chain in which its bonds are much relaxed. As a result energy in the form of heat is given off to the surroundings, and the disorder of the surroundings increases. The total system, consisting of both beakers, is also disordered because in the transfer the impure liquid B has gained in volume at the expense of the pure liquid A.

Let us assume that this kind of transfer is repeated many times. In due course the chain molecules in beaker B becomes so long that they start to undergo bending strains as they try to occupy the

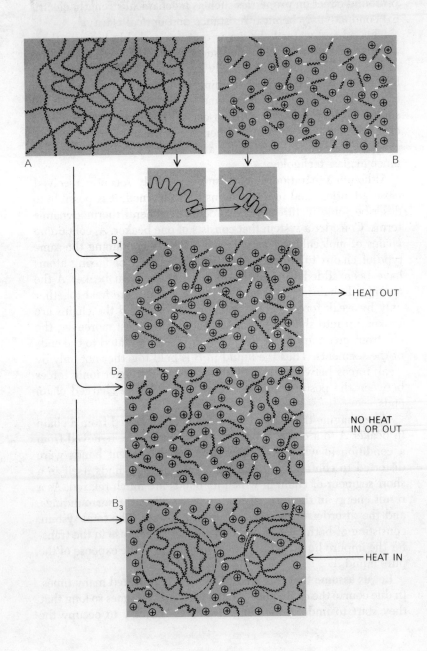

space between fluxing ions more efficiently. This means that as each additional molecule is transferred from A to B the heat evolved becomes less and less. Finally a point is reached where the relaxation energy given up in the transfer of a molecule of A will just balance the ionic bond energy taken up by the incremental separation of fluxing ions and chain ends. At this point no heat at all is evolved, and the disorder that results consists solely of the disorder due to mixing pure molecules with impure solution. Since the system and the surroundings taken together experience a net increase in disorder, the transfer process is still thermodynamically possible.

The next molecule transferred, however, will require external energy if it is to be incorporated into a chain segment in B. This energy must come from the surroundings as heat, and it is utilized partly in distorting the covalent bonds of the lengthening B chains and partly in separating chain ends and flux ions. Finally a state may be reached in which the ordering taking place in the surroundings, because of the loss of heat to beaker B, may become equal to the disordering resulting from the further mixing of A and B. Such a condition may arise easily if the attractive forces between chain ends and flux ions are fairly strong and if the temperature is fairly low. Once this condition is reached something of a dilemma is presented: on the addition of further A molecules to B, a net increase in the ordering of the "universe" would tend to occur, and this, as far as we know, cannot happen. The difficulty is resolved, however, because the last molecules transferred simply gather together in local regions, partially exclude fluxing ions so that the attractive forces in these local regions are reduced, and pack themselves as long chains in an open structure resem-

Two-liquid glass structure, like that shown in the photograph on page 82, is believed to arise through a process similar to that schematized here. A is a melt of pure glass made up of long, tangled chains. B is a fluxed glass in which the positive charges of flux ions are matched by the number of electrons (white dashes) at the ends of short, straight chains. The sequence B_1, B_2, B_3 shows what happens if A chains are slowly transferred to a beaker of B chains. As the detail shows, the short B chains lengthen by repeated transfer of molecules from the longer A chains. At first there is an outflow of heat. Later the flow is reversed, but thermodynamic considerations indicate that the melt will then separate into two phases (B_3).

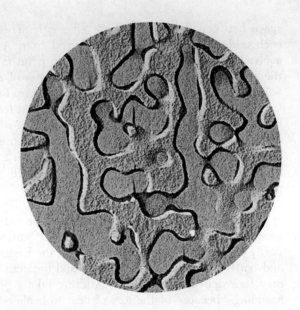

The two-phase glass shown in this electron micrograph is a sodium borate glass that has separated on cooling into two continuous and interpenetrating phases, each of them a glass. The two phases differ markedly in chemical composition and properties. The character of the resulting product depends not only on these differences but also on the manner in which one phase is distributed in the other. The micrograph was made by the author.

bling their original configuration in A. In this fashion separation of one liquid into two liquids with different properties overcomes the impending dilemma and provides an avenue for continued mixing.

Numerous two-phase structures are possible in glasses [see illustration above] and are present in many commercial forms of glass. Recently it has been found that a prior two-liquid formation may play a significant role in determining the path of crystallization and the final properties of "crystallizable" glasses. These are glasses that are first melted and formed in a desired shape and then, by prolonged heat treatment, converted into a strong and durable ceramic that is part glass and part crystal. Ceramics of this type, with such trade names as Pyroceram, Re-X and CER-VIT, are finding increasing use in industry and in the home.

One of my colleagues maintains that glasses are the next best thing to universal solvents. His point is well taken; practically all elements can be dissolved in glasses to a considerable degree. In addition, many kinds of chemical reaction—decomposition, precipitation, dissolution, ion-exchange, oxidation and reduction—can be carried out directly inside a glass. There is the added advantage that such reactions can often be halted, or frozen in, at any stage desired.

Control over such processes has made it possible to produce glasses that show laser action, lighten or darken in response to light, exhibit semiconduction and photoconduction, fluoresce and transport ions selectively. There is even a hope that a glass may eventually demonstrate superconductivity. As glass science continues to provide a rational basis for the glassmaker's art many new and unexpected phenomena will certainly be encountered.

In synthesizing long-chain molecules man imitates natural polymers such as cellulose. Today nature is being outdone, and polymers are evolving that may be rigid enough to serve for heavy construction.

HERMAN F. MARK

The nature of polymeric materials

Life depends fundamentally on organic polymers. They provide not only food but also clothing, shelter and transportation. Indeed, nearly all the material needs of man can be supplied, and in many times and places they have been, by natural organic products. The list of these materials and things made of them is very long: wood, fur, leather, wool, flax, cotton, silk, rubber, oils, rope, sails, paper, parchment, canvas, paint, stringed musical instruments, bows, arrows, tents, houses, ships and shoes and sealing wax.

The natural organic polymers from which such things are made include proteins, cellulose, starch, resins, lignin and a few other classes of compounds. Because of the complexity and chemical fragility of their molecules, the natural organic polymers, although

Gross structure of a polymer can be partly crystalline and partly amorphous, as shown in this photomicrograph. The material is a silicon polymer that has been melted, resolidified and photographed under polarized light. The radially symmetrical arrays emanating from light centers are spherulites, or crystalline structures; other parts of the polymer show amorphous domains. The material was photographed at the General Electric Research and Development Center under the direction of F. P. Price.

known and used for ages, defied attempts to analyze their molecular structure until very recently. Since 1920 modern methods of physical and chemical analysis have uncovered the principles that govern the properties of the natural polymers, and organic chemists, going to school to nature, have created a new industry of man-made organic polymers. It has become a major enterprise in all industrialized countries; in the U.S. more than 13½ billion pounds of synthetic plastics and resins, totaling more than $6 billion in value, were produced in 1966.

The principal products are fibers, packaging materials, synthetic rubbers, coatings, adhesives and a galaxy of the materials called "plastics." The markets for these products are now approaching saturation, and the industry is therefore looking for new worlds to conquer. The most inviting prospect for large-scale expansion of the use of synthetic polymers lies in the field of construction materials—not only for houses and other buildings but also for automobiles, airplanes and boats. Plastics and synthetic coatings are already in common use in the finishing sector of building construction, as floor tiles, insulation, trim and so on. The new field now envisioned by chemists is the creation of organic polymers that will serve in primary structural capacities: as load-bearing members, structural shells, walls and conveyors of various utilities.

Synthetic polymers now available already possess several of the properties required in a structural material. They are light in weight, easily transported, easily installed, easily repaired, highly resistant to corrosion and solvents and satisfactorily resistant to moisture. They fall short of the requirements in structural strength, long-life durability and resistance to high temperature. The outlook is good, however, for the development of polymers that will meet these demands. The chief question has to do with their cost: it remains to be seen if synthetic polymers can be made inexpensive enough to compete with the traditional structural materials—metals and ceramics.

It may seem odd that man came so late to the investigation of organic polymers. As the principal means of supporting life, the natural polymers—proteins, cellulose and so on—dominated his existence, and even in ancient times people developed great ingenuity, craftsmanship and sophistication in the handling and use of these materials. Yet as late as the end of the 19th century, when the science of organic chemistry excited high enthusiasm and fired many of the keenest minds, polymer chemistry got little attention. Chemists attacked sugar, glycerol, fatty acids, alcohol, gasoline and other ordinary organic compounds—dissolving, precipitating, crystallizing and distilling to learn what these substances were composed of and how they were put together. Only feeble and ineffectual efforts were made, however, to investigate such common materials as wood, starch, wool and silk. The substances composing these materials could not be crystallized from solution, nor could they be isolated by distillation without being decomposed.

It remained for some powerful physical instruments of the 20th century—the ultracentrifuge, electron microscope, viscometer, osmometer, diffusion cell and X-ray diffraction apparatus—to reveal the polymers in all their intricacy. Their molecules were discovered to be almost incredibly large: the molecular weights ran as high as millions of units, whereas simple organic substances such as sugar and gasoline have molecular weights in the range of only about 50 to 500. The giant molecules turned out to be composed of a large number of repeating units; they were consequently given the name "polymer," from the Greek words *poly* ("many") and *meros* ("parts"), and the building blocks were called monomers. Most polymers were found to have the form of long, flexible chains. Packed side by side in a bundle, the molecules formed a regular array with a crystalline structure. Examination of natural polymeric materials showed that their structure was complex—partly crystalline and partly amorphous. If the crystalline structure predominated, the material was relatively strong, rigid and resistant to heat and dissolution; if the structure was amorphous, the substance was soft, elastic, absorptive and permeable to fluids.

Having learned these ground rules, chemists undertook to synthesize artificial polymers. They soon succeeded in forming new monomers from simple and inexpensive raw materials, in stringing the monomers together efficiently into long chains, in obtaining quantitative data on the molecular weight and structure of polymers and in ascertaining how structural details influence a polymer's properties. These efforts, beginning in 1920, had led by 1940 to the establishment of industries producing synthetic fibers and numerous other polymeric materials, many of which were less expensive and superior in various ways to the natural materials after which they were modeled.

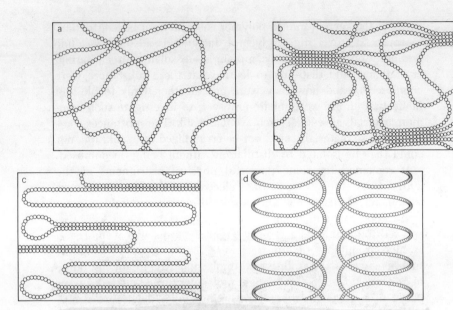

Systems of polymer chains include amorphous arrays (a) *and various forms of crystallization. There can be crystallization between chains* (b) *and also chain-folding, which is crystallization of a chain on itself. Chain-folding can take a laminar form* (c) *or a helical form* (d); *the helical form can be either left-handed or right-handed. A predominantly amorphous polymer is soft, elastic and permeable to fluids; a polymer with a predominantly crystalline structure is appreciably stronger and more rigid.*

Looking ahead now, what might be done to improve the strength of organic polymers and enlarge their realms of usefulness? The properties that must be built into them to qualify them as structural materials (for buildings and vehicles) can be specified in detail: (1) rigidity sufficient to bear a load of at least 700,000 pounds per square inch; (2) a tensile strength of at least 100,000 pounds per square inch; (3) an elasticity of at least 10 percent (to resist breaking or tearing); (4) a melting or softening point above 500 degrees C.; (5) high resistance to damage by heat, radiation and corrosive chemicals; (6) high resistance to the action of solvents and swelling agents, even at elevated temperatures.

Let us examine the characteristics of polymeric structure and consider how they can be exploited to strengthen the properties of such a compound. One characteristic is the capacity of the chains to organize themselves in a crystalline structure. When a polymer with a regular architecture (that is, a highly ordered arrangement of the atoms or atomic groups in the chain) is subjected to a mechanical treatment that orients the chains, they have a strong tendency to line up in parallel and form crystallites. The individual bonds connecting the chains are not strong, but their large numbers and regular spacing give considerable rigidity to the structure, hence the material becomes hard, insoluble and resistant to softening by heat. Thus even crystallized polyethylene, whose chains are held together only by weak van der Waals forces, is a strong, tough, abrasion-resisting material with a fairly high melting point (130 degrees C.). Similarly, polypropylene, another polymer with a regular chain architecture, is rigid and has a melting point of 175 degrees in the crystallized form, and polystyrene is still more resistant to softening, with a melting point around 230 degrees. Even stronger and more heat-resistant materials can be obtained by crystallizing polymers that contain polar groups (groups in which there is a separation of positive and negative electric charges), as these provide stronger interchain bonding. Examples of such materials are nylon, Saran, Dacron and Mylar.

Crystallization is one of two principles that have long been applied to give strength and resistance to polymers; the other principle is chemical cross-linking of the chains. A substance such as sulfur is added to the polymer for the purpose of forming strong chemical bonds between the chains. The process is quite different from crystallization. Whereas crystallization is a physical phenomenon that depends on orientation of the chains, is not influenced by temperature and can be reversed without decomposing the polymer, cross-linking is an effect that depends on a chemical reaction rather than on physical orientation of the chains, is strongly accelerated by elevating the temperature and is not reversible; because of the strong and randomly located bonds connecting the chains the material is not merely softened but breaks down altogether at temperatures high enough to melt it.

A good example of a cross-linked polymer is rubber. In its most common form (the vulcanized rubber of tires) the chains are linked by a certain quota of sulfur atoms that leaves the rubber elastic. By adding more cross-links one can progressively stiffen the piece until it becomes the hard substance called ebonite—a material that is very rigid, has an extremely high softening temperature and is completely insoluble and unswellable.

Ebonite, one of the oldest of the "plastics" (actually it is not a plastic but a thermosetting material), is a prototype of many other cross-linked polymers. These include various hard rubbers, formaldehyde-based products, polyesters and thermoset resins that are hardened by grafting styrene onto a polyester backbone containing aliphatic double bonds.

If crystallization and cross-linking were the only available means of strengthening and toughening organic polymers, there would be little hope of synthesizing a polymer that would fill the bill as a structural material. Apparently the most that could be achieved by crystallization and cross-linking, even in combination, is a rigidity modulus of around 450,000 pounds per square inch and a melting temperature of about 350 degrees C.—far short of the specifications required for a building material. There is, however, a third principle, now under active investigation, that holds high promise.

The materials I have been discussing are all composed of inherently flexible chains. The segments of these chains are linked by rotatable bonds and can easily bend, kink or fold on one another (like the jackknifing of a tractor-trailer). Hence the only way to achieve stiffness and strength is to pack the chains together in rigid assemblies, either through crystallization or cross-linking. Obviously the rigidity could be enhanced by assembling chains that were themselves intrinsically stiff.

There are several possible ways to stiffen a polymeric chain. One is to hang bulky groups of atoms on the chain to restrict bending. This principle is exemplified by polystyrene. In the polystyrene molecule benzene rings are attached to the carbon backbone of the chain. The consequent stiffening is sufficient to make polystyrene a hard plastic with a fairly high softening point (90 degrees C.), even though its chains are not cross-linked or packed in a crystalline array. The absence of crystallinity makes the material completely transparent, and the absence of cross-linking makes it readily moldable. Another example is polymethylmethacrylate (Lucite), which, by virtue of the methyl (CH_3) and methacrylate ($COOCH_3$) groups attached to carbon atoms along the chain, is a hard, brilliantly transparent material with a softening point of 95 degrees.

Materials stiffened by attaching bulky groups to the chain have a weakness, however: they are fairly easy to dissolve and are subject to swelling. It seems that the bulky groups allow ready penetration of the system by solvents and swelling agents. Because of this behavior, research attention is focused on ways to stiffen the backbone of the chain itself.

Cellulose, the structural framework of wood, is a classic example of a polymer with an intrinsically stiff backbone. Its chain molecule is a string of condensed glucose (sugar) molecules, which are ring-shaped. Building on this chain, chemists long ago produced cellulose acetate and cellulose nitrate, which are hard, transparent, high-melting and amorphous thermoplastic resins. In these compounds the acetate ($OCOCH_3$) and nitrate (ONO_2) groups are attached irregularly along the chain and account for the absence of crystallinity. The attached groups tend, of course, to give the compounds the weaknesses I have mentioned: cellulose acetate is rather sensitive to the action of solvents and swelling agents, and cellulose nitrate, although less sensitive, is still dissolvable by some solvents.

Chemists are now experimenting with other monomers that, like glucose, contain ring-shaped groups of atoms and therefore have intrinsic rigidity. With these monomers they have synthesized a number of hard polymers endowed with strong properties: some of the materials can be exposed to temperatures up to 500 degrees C. for long periods without softening or deterioration and are completely insoluble in all organic solvents at temperatures up to 300 degrees.

One promising approach is based on manipulation of the polyphenylene chains. The monomer, phenylene, comes from benzene and has the ring structure. In the chain the phenylene rings are linked by carbon-carbon single bonds that do not allow sufficient rotation to kink or bend the chain, and as a result the chain cannot fold even at rather high temperatures. The polyphenylenes are rigid and high-melting, have a pronounced tendency to crystallize and are highly insoluble. So far, however, the inherent stiffness of the phenylene chain has not been fully capitalized on, because no way has yet been found to build up its polymers into sufficiently long chains.

Another interesting group of polymers based on aromatic chains (chains made up of benzene-type rings) are the "ladder polymers." Produced by a series of stages involving the progressive heating of an unsaturated precursor (typically polyacrylonitrile or 1,2-polybutadiene), these polymers correspond to a "chain" of graphite, with one carbon atom in each ring replaced by nitrogen. The materials are hard and completely insoluble and unmeltable.

We see, then, that there are many possibilities for synthesizing long, stiff chains, and that the resulting polymers possess the valuable properties predicted for this type of construction. The next question is: Can we obtain even better results by combining two or more of the three principles (crystallization, cross-linking and stiff chains)?

Let us summarize what has already been accomplished by each of the single approaches and by combinations of two principles. The crystallization method has produced a large number of thermoplastic materials, particularly fibers and films; they include polyethylene, polypropylene, polyoxymethylene, polyvinyl alcohol, polyvinyl chloride, polyvinylidene chloride and polyamides such as 6 nylon and 66 nylon. The cross-linking approach has yielded the hard rubbers, thermosetting resins, polyesters, network polyepoxides, polyurethanes and the resins and plastics formed by compounding formaldehyde with urea, melamine or phenol. In the group of polymers based on stiff chains are polystyrene, polystyrene derivatives, chain polyepoxides, polymethylmethacrylate, polycarbonates, polyesters, polyethers and other products.

How much can be gained by combining the stiffening techniques? Consider first the combination of crystallization and cross-linking. The principal materials produced by this combination are the crystallizable rubbers: natural rubber, 1,4-cis-polybutadiene, polyisoprene and neoprene. If a great many cross-links are introduced, the polymer loses its crystallinity and becomes amorphous; ebonite is a case in point. To add stiffness to rubbers and other elastomers that do not crystallize readily, a solid filler is usually introduced. The particles of the filler (a hard, finely divided material such as carbon black, silica or alumina) attach themselves strongly to the polymer chains by adsorption and serve to stiffen the chains by immobilizing the segments; in effect they produce a kind of crystallization of the system.

Crystallinity and cross-linking in combination can raise the rigidity of organic polymers to a modulus of 450,000 pounds per square inch and the melting temperature to 350 degrees C., as I have already mentioned, but that seems to be about the limit.

What if we combine crystallization with the use of intrinsically stiff chains? Again the combination results in substantial gains. Cellulose provides a good illustration. Its rigid chains give a material high tensile strength and a high melting point even at a relatively low degree of crystallinity. With enhancement of the chain stiffness and the crystallinity, compounds based on the cellulose backbone can be built up to polymers that are extremely rigid, do not melt at all and are soluble only in a very small number of particularly potent solvents. Cellulose's excellent potentialities for forming tough fibers and films have given rise to many spectacularly successful products. One example of an effective merger of chain stiffness and crystallinity is cellulose triacetate: the polymer's capacity to crystallize endows it with several advantages over ordinary cellulose acetate, notably greater resistance to organic solvents and improved thermosetting characteristics.

Terylene is another illustration of the efficacy of combining the two principles. In this crystalline polymer (chemically described as polyethylene glycol terephthalate) the chains are only moderately stiff and are held together by extremely weak lateral forces because no hydrogen bonds are available. Even so, the combination of crystallization and chain stiffness suffices to give the fiber high strength and the high melting point of 260 degrees C.

The third possible pairing of principles—the combination of chain stiffness and cross-linking—has also been explored, and this too has yielded encouraging results. For example, stiff-chain epoxy polymers have been "cured" to greater rigidity and resistance to softening by building up the number of cross-links between the chains.

Since combinations of two of the three strengthening principles have proved effective in enhancing polymer properties, chemists are naturally hopeful of achieving even better results by combining all three principles [see diagram on the next page]. Much exploratory work is now being done along that line, and already certain interesting successes have been obtained. One of these is the application of the three principles to improve the properties

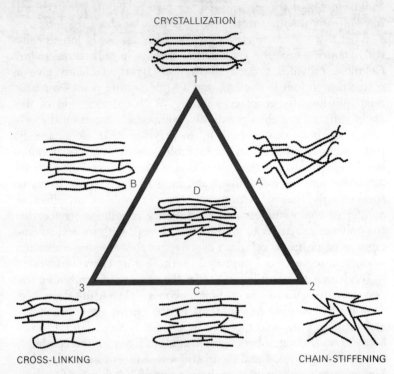

Combined features make it possible to achieve various properties with polymers. Each corner of the triangle represents one of the three basic principles for making a polymer rigid and temperature resistant; the sides and the center of the triangle indicate various combinations of the principles. The chart on the facing page gives examples of the possibilities.

of cotton and rayon. The fiber is given two treatments: a mechanical treatment to crystallize the stiff cellulose chains and treatment with a chemical agent that introduces cross-links. It has been found that the cross-linking substantially improves the recovery power and wrinkle-resistance of fabrics made from such fibers, without diminishing their other desirable properties. Similarly, promising results have been obtained in strengthening polymers

LOCATION	POLYMER CHARACTERISTICS	EXAMPLES	USES
1	Flexible and crystallizable chains	POLYETHYLENE	Pails, pipes, thin films
		POLYPROPYLENE	Steering wheels
		POLYVINYL CHLORIDE	Plastic pipes and sidings
		NYLON	Stockings, shirts, dresses, coats
2	Cross-linked, amorphous networks of flexible chains	PHENOL-FORMALDEHYDE	Television casings, Telephone receivers
		CURED RUBBER	Tires, transport belts, hoses
		STYRENATED POLYESTER	Finish on automobiles and appliances
3	Rigid chains	POLYMIDES	High-temperature insulation
		LADDER MOLECULES	Heat shields
A	Crystalline domains in a viscous network	TERYLENE (DACRON)	Fibers and films
		CELLULOSE ACETATE	Fibers and films
B	Moderate cross-linking with some crystallinity	NEOPRENE	Oil-resistant rubber goods
		POLYISOPRENE	Particularly resilient rubber goods
C	Rigid chains, partly cross-linked	HEAT-RESISTANT MATERIALS	Jet and rocket engines and plasma technology
D	Crystalline domains with rigid chains between them and cross-linking between chains	MATERIALS OF HIGH STRENGTH AND TEMPERATURE RESISTANCE	Buildings and vehicles

of the epoxy and urethane types. In these systems, starting with stiff chains that have been cross-linked, the experimenters add fillers to produce the equivalent of crystallization.

There is good reason to expect that thorough and systematic exploration of the new threefold attack will lead to many new and interesting organic polymers, perhaps even to superior materials for building our houses and vehicles.

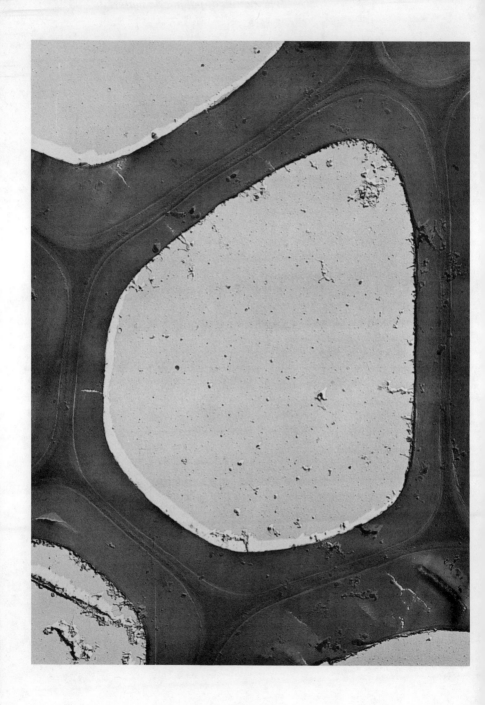

Metals, ceramics, glasses and polymers can be combined in materials that have unique properties of their own. Nature uses this principle in wood and bone; man applies it in a new family of supermaterials.

ANTHONY KELLY

The nature of composite materials

Until quite recently a composite material was regarded simply as a combination of two materials. It is still that, but in modern technology it has acquired a broader significance: the combination has its own distinctive properties. In terms of strength or resistance to heat or some other desirable quality it is better than either of the components alone or radically different from either of them.

This concept of composite materials is leading to the design and manufacture of a new range of structural materials that may bring about far-reaching changes in engineering and construction. The principal attraction of composite materials is that they are lighter, stiffer and stronger than anything previously made. At present they are used mainly to meet the severe demands of

Wood is a natural composite, consisting of cellulose fibers in a matrix of lignin. In this electron micrograph, which was made at the New York State College of Forestry at Syracuse University, a cross section of aspen is enlarged approximately 11,000 diameters. The darkest material is lignin, the slightly less dark material the wall of a cellulosic cell, and the pale central object the lumen, or cell cavity.

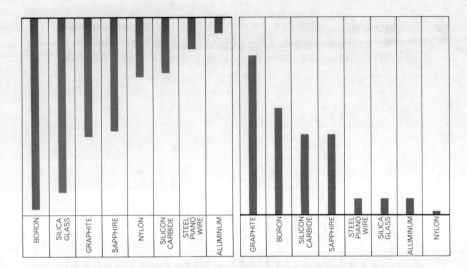

Strength and stiffness of various materials are compared. Strength (left) is represented as greatest free-hanging length; that of boron is 189.4 miles. Stiffnes (right) is represented on an arbitrary scale indicating relative stiffness per unit weight.

supersonic flight, the exploration of space and ventures into the deep waters of the oceans. Someday they will be put to less exotic uses, such as the construction of buildings. When that happens, one can expect to see square corners and heavy masonry give way to airy domes and delicate tracery.

Like synthetic polymers, modern composite materials imitate nature. Wood, for example, is a composite consisting of cellulose and lignin. Cellulose fibers are strong in tension but flexible; lignin acts to cement them together and endow the material with stiffness. Bone is a composite of the strong but soft protein collagen and the hard but brittle mineral apatite. Modern composites achieve similar results even more effectively by combining the strong fibers of a material such as carbon in a soft matrix such as epoxy resin. A more familiar example is fiber glass, which consists of glass fibers in resin.

Such materials go well beyond the older concept of composites, in which two or more materials were combined to rectify some shortcoming of a particularly useful component. For example, early cannons, which had barrels made of wood, were bound with brass because a hollow cylinder of wood easily bursts under internal pressure. By the same token, ordinary steel is normally covered with paint because of steel's strong affinity for oxygen. In this older sense nearly all engineering materials are composites of some kind.

An elementary example of the newer concept of composites is the bimetallic strip used in such devices as the thermostat. The strip might be made of a flat piece of brass and a similar piece of iron. If the two pieces were separate and were heated simultaneously, the brass would expand more than the iron. If the two pieces are welded together and the composite is heated, the greater expansion of the brass forces the iron to bend, and the bending of the iron forces the brass to bend. The bending can be used to indicate temperature or to activate an on-off switch.

This example illustrates two points about modern composites. The first is that either material alone would be useless in the application; the combination has an entirely new property. The second is that the two components act together to equalize their different strains. This behavior, called combined action, is most important in the design of composite materials, as we shall see.

For an engineer designing a structure such as a supersonic aircraft an ideal load-bearing member is made of something that is very stiff, very light, very strong in tension and not easily corroded; that expands very little with changes in temperature, and that has a high resistance to abrasion and a high melting point. Because stiffness and lightness and strength and lightness are required together in contemporary applications, it is the stiffness per unit weight and the strength per unit weight that must always be considered. The manner in which several materials meet these considerations under laboratory testing is set forth in the chart on the facing page. One sees there that the strong materials that are also light and stiff are ceramics: materials such as glass, graphite, sapphire, Carborundum and boron. Metals, which one usually associates with high strength, do quite poorly by comparison. Even the strongest steel—itself more than seven times stronger than normal steel—is an unimpressive material on a unit-weight basis. The resins and polymers do well in terms of strength to weight but badly (much worse than metals) in terms of stiffness to weight.

Glass, Carborundum and graphite not only are superior to the common metals in strength and stiffness per unit weight but also have high melting points and do not expand significantly when they are heated. Moreover, they can be made from inexpensive raw materials such as sand, coke and coal. Why, then, have they been so little used as structural materials?

The reason, which held until the discovery of the composite principle, is that their high strength is realized only under rather special conditions. The most important condition is that the specimen have no internal cracks and have a very smooth surface that is also free from cracks, small notches and steps. The significance of cracks and surface dimples can be demonstrated by a comparison of three common materials—strong steel, aluminum and glass—for their resistance to the propagation of cracks.

Using as a test the size of crack that would cause immediate breakage of a sheet of the material if it were stressed to 100,000 pounds per square inch, one finds enormous differences. Strong steel can tolerate cracks or machined notches or other sharp surface irregularities up to an inch in depth. In aluminum the crack can be no deeper than 1/64 inch. Glass breaks if the crack is deeper than 1/10,000 inch.

The measure of a material's ability to retain its strength in the presence of cracks is determined by what is called the work of fracture of the material, which is to say the energy required to break it. Glass has a very small work of fracture and well-designed strong steel a very high one. The inherently strong materials such as silicon carbide, boron and graphite all behave somewhat like glass: their work of fracture is small and so they are quite vulnerable to the presence of cracks.

Metals are normally used to bear large stresses because they can accommodate themselves to cracks. With a metal the engineer obtains high strength without having to take extreme care to eliminate all but the tiniest cracks. Polymeric materials such as polyethylene are rather like metals in resistance to cracks, although they will not take as much stress as metals.

The basic chemical reason metals and polymers are so much more resistant to cracks than ceramics are is that the interatomic forces in metals and the intermolecular forces in polymers do not depend on a particular directional alignment to achieve strength.

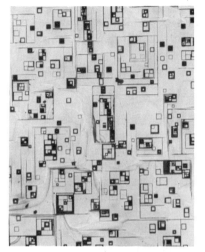

Surface defects on crystals of magnesium oxide indicate the brittleness of most ceramic crystals. The defects in the crystal at left were made by heating the crystal in air; the defects in the crystal below were caused by dropping small particles of dust onto the surface. The ease with which most ceramic crystals crack is the reason ceramics are so little used for heavy structural work in spite of their great strength. In a composite of ceramic fibers in a matrix the soft matrix prevents the propagation of any cracks.

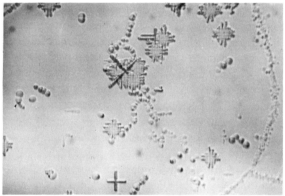

Moreover, the chemical bonds of metals and polymers are unsaturated, that is, the atoms or molecules of such materials are readily capable of forming new bonds. Ceramics, on the other hand, have highly oriented forces and saturated bonds. In metals and polymers atoms or molecules always slide over one another at the leading edge of a crack. The crack therefore cannot penetrate the internal structure of a metal or polymer as easily as it can the structure of a ceramic; a much larger stress is needed to make a crack run through a metal or a polymer and divide it into two pieces.

The familiar forms of ceramics are represented by glass, chalk and sand; by abrasive powders such as corundum and Carborundum, and by gems such as diamond and ruby. It is because ceramic crystals are so vulnerable to cracks that ceramic materials are often used in the form of a powder, to which they are easily reduced. Even as a powder they scratch steel because in fact they are stronger than steel.

In sum, if a ceramic material is unscratched, it can be very strong. If it is flawed in any way, however, it breaks easily. The lack of resistance to cracks is called brittleness. Ceramics are usually fragile because ceramic crystals are almost always marred by cracks or surface irregularities. Even if they are not, such imperfections can be introduced all too readily [*see illustration on page 101*].

In order to use a ceramic in a modern composite material it is necessary to divide it into small pieces, so that any cracks present cannot find a continuous path through the material, and to bind the pieces together in a matrix. The ceramic is often put into the composite in the form of fibers. The properties of the matrix are of vital importance. First, it must not damage the fibers by scratching them, which would introduce cracks. Second, it must act as the medium by which stress is transmitted to the fibers; it should be plastic and adhesive so that it holds the fibers much as deep, soft mud holds one's leg when one steps into it. Third, the matrix must deflect and control cracks in the composite itself.

All the required mechanical properties of a matrix are met by a polymer or a metal such as aluminum or copper. Such a matrix is soft, which is to say weak in shear, and so it does not scratch the fibers. To carry out its other two functions the matrix needs qualities that are somewhat at odds with one another and so call for compromise.

To perceive how the matrix performs those functions the reader must bear in mind the difference between stress and strain. Stress is the externally applied force; it is usually measured in pounds per square inch. Strain is the distortion of the material when it is under stress; it is expressed in terms of the change in the original shape of the material.

Let us now consider more closely the function of the matrix in transmitting stress. The composites of the highest strength contain

The nature of composite materials 103

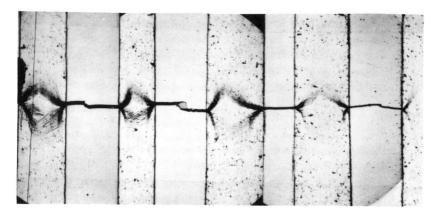

Effect of matrix in blunting cracks is shown in a composite of tungsten fibers in a copper matrix. The tungsten fibers, which are the clearer bars, broke when the material was pulled in a direction parallel to the fibers; the copper did not. The dark flow lines in the copper show that the matrix has retarded the cracking by the process of shear. The movement of the crack was from left to right.

aligned fibers. If such a composite is stretched parallel with the fibers, the principle of combined action comes into play, so that the strains in the fiber and in the matrix are virtually equal. One chooses a matrix that yields or flows in a plastic manner, so that when the fibers and the matrix are under equal strain the stress within the fibers is enormously greater than it is in the matrix. The difference is so pronounced that in working out the breaking strength of the composite the contribution of the matrix can be regarded as negligible.

When the fibers are highly stressed, some that have cracks will break. The beauty of a composite material is that such a crack will usually be unimportant. A close look at the leading edge of a crack in such a composite shows that the propagation of the crack through the brittle reinforcing material is hindered by the softness of the matrix [*see illustration above*].

Two other effects prevent cracks from running through a composite material. The first is that, although the reinforcing fibers may fail, they do not all do so in one plane. To make a crack extend all the way through the material it would therefore be necessary to pull the fibers out of the matrix one by one as they broke. Work must be done by the applied stress in pulling out the fibers against the holding force of the matrix, and so the resistance to crack propagation is increased. The "pull out" work, which makes a large contribution to the work of fracture in composites consisting of brittle fibers in resin, is a true composite property: it cannot be attributed to either component alone.

The second crack-controlling effect is achieved by regulating the degree of adhesion between the fibers and the matrix. If the adhesion is low, the material is weak in a direction at right angles to the fibers. This is an advantage, however, if a crack starts to run at right angles to the fibers; the crack is deflected along the weak interface and rendered harmless as far as the desirable properties parallel to the fibers are concerned.

So far we have been dealing with the strength and crack-controlling properties of a composite material being pulled in a direction parallel to the fibers, that is, when the material is under tension. When the material is under compression parallel to the fibers, the specimen breaks down by buckling and shear. Here a need arises for one of the compromises to which I have referred.

Under compressive loading the stiffness of the fibers should be at a maximum to resist buckling, and the interface between the fiber and the matrix should have a high tensile strength in order to resist splitting. For the material to resist cracking under tension, however, a weak interface is required, so for resistance to both tension and compression a compromise must be made.

Compromise is also required when a composite is pulled in tension at an angle to the fibers, that is, when it is subjected to shear. Under shear, as under compression, the composite is less strong than when it is under tension. In other words, the strength of a composite tends to be highly directional, just as wood is strong parallel to the grain and weak at right angles to it. The compromise that is applied to composites to counteract their weakness under compression and shear is one that has long been used with wood, namely lamination, as in plywood. In some modern composites variously aligned layers are stuck together to provide strength in a number of directions. A price has to be paid: the laminated material is weaker in any particular direction than it would be in one direction if all the fibers were aligned.

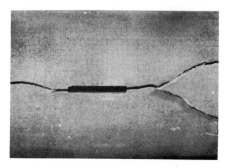

Laminated steel can retard cracks. When pressure vessels were made to burst, one of standard steel (top) developed longer and larger cracks than one made of laminated steel (bottom), which consisted of steel sheets bonded together. Slots were made before testing as built-in defects and were then sealed on the inside so that the vessels could be burst.

Crack-controlling properties and lamination have recently been applied in conventional metallurgy. Steel can be made more resistant to cracks by cementing together thin sheets of it with soft solder. This laminated steel is, of course, a composite material.

When fibers of an extremely brittle material such as glass are used in a composite, they will always be marred by some flaws. When such a composite material is stressed, some of the fibers break before others. Obviously the part of a broken fiber close to the break will not support any load. A short distance from the break, however, the unbroken part of that same fiber will be carrying as much load as the surrounding unbroken fibers. The reason is that when the fiber breaks, the two ends attempt to pull away from each other but are prevented from doing so by the matrix, which adheres to the fiber. As the fiber attempts to relax, the flow of the matrix parallel to the stress counteracts the tendency. Shear forces come into play and gradually build stress back into the broken fiber [*see illustration on the next page*].

The fact that the matrix in a composite builds stress into broken

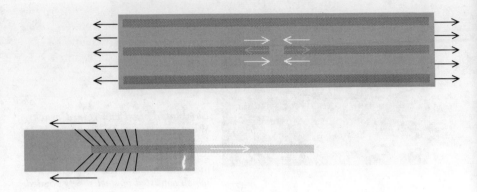

Broken fiber in a fiber-reinforced composite causes little damage. The reason is depicted schematically at top in a representation of a broken fiber and two unbroken ones in a matrix. When the central fiber breaks, with the material stressed as shown by the solid arrows, the two pieces of fiber attempt to pull away from each other but are prevented from doing so by the shear forces (black arrows) in the adhering matrix. Forces at work on a broken end of a fiber are represented in more detail in the illustration at bottom.

fibers means that the principle of combined action would still be realized even if all the fibers were broken. Such an effect is achieved with large numbers of small fibers. Composites can accordingly be made with short lengths of fiber, none of which needs to run through the entire piece of material.

This fact has two substantial advantages for the application of the principle of fiber reinforcement. Since small lengths of fiber can be used, pieces of material can be built up (and given multidirectional strength) with layers consisting of matrix and short fibers. The second advantage is a pure bonus. The strongest materials known are the short, single-crystal filaments known as whiskers. Because a fiber-reinforced composite does not demand continuous fibers, whiskers can be used. Although no whisker-reinforced material has yet been employed in a practical structure, there is much interest in the manufacture of whiskers of such materials as silicon carbide, boron carbide and aluminum oxide, which, being ceramics, are very stiff. Composites reinforced with whiskers have been made in various laboratories and have demonstrated remarkable properties.

The largest tonnage of composite materials now being manufactured is in the form of fiber glass. Glass, because of the manner in which silica melts, is easily drawn from the molten state into thin, high-strength fibers. Drawn glass is vulnerable to attack by water, which drastically reduces its strength, and the fibers have to be given a protective coating.

Fibers so prepared can readily be put into a matrix of unsaturated polyester resin. The resin is originally in liquid form, and the reaction to make it set around the fibers can be promoted at low temperature and low pressure. This means that the glass fiber is not grossly damaged in the process of forming a composite. The fact that the fibers remain largely intact is one of the chief advantages of glass-reinforced plastics. Large pieces of such material can be built up easily by the application of successive layers of resin and glass. Since any required shape can be built up piece by piece, there is in principle no limit to the size of object that can be constructed.

In another technique large vessels to contain gases at high pressure are made with continuous filaments of glass fiber. The filaments are wound onto a mandrel after passing through a bath of liquid resin. The pattern of winding can be controlled so as to put more fibers along the directions to be highly stressed, and the fibers can even be stretched. These methods of fabrication are completely different from conventional metallurgical techniques for making strong alloys and have the enormous advantage of not calling for high temperatures or high pressures.

Glass-reinforced plastics do have some significant disadvantages. One is that glass fiber, although it is very strong, is not very stiff. Bridges and airplane wings cannot be made out of fiber glass because the fibers can stretch, and the material will bend too much under load. Secondly, resins burn, char or flow at temperatures of around 200 degrees centigrade. This disadvantage can be somewhat offset in the most modern materials by the use of high-temperature polyimide resins reinforced with fibers of pure silica glass. Such composites have withstood temperatures of more than 300 degrees C. for several days. There is a limit to the temperature resistance of glass-reinforced plastics, however, because the glass itself does not remain strong at temperatures much in excess of 400 degrees.

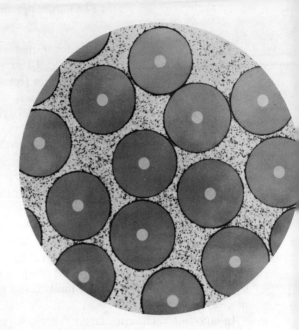

A man-made composite, produced at United Aircraft Research Laboratories, consists of boron fibers in aluminum. Fibers (dark circles) are each .004 inch in diameter. The lighter circle in each of the fibers was made by the wire on which the boron was formed.

Fortunately the stiff materials held together by covalent bonds—boron, carbon and others—have high melting points. In fact, stiffness and a high melting point go together with covalent bonds because of the large amount of energy needed to break such a bond. Therefore the materials that replace glass fiber because of their greater stiffness should also overcome the temperature limitations of glass.

Very stiff fibers of graphite, boron and silicon carbide have been made successfully. At high temperatures these materials often evaporate instead of melting. Special methods have therefore been used to produce the fibers. Graphite fibers, for example, are made by controlled heating of synthetic fibers of the type used in artificial textiles; the heating decomposes the material but the carbon is retained and turned into graphite. The production of boron fibers proceeds essentially by the decomposition of boron chloride or bromide on a hot filament to form a tube of strong boron over the filament. Both tungsten wire and silica coated with graphite (to make it conducting) can be used as the initial filament.

The fibers of graphite and boron now in production are more than twice as stiff as steel. Since they are less than a third as dense as steel, when they are put into a resin matrix they make a composite with a stiffness per unit weight much higher than steel's. The fibers are also very strong, and so the strength of steel is also exceeded on a unit-weight basis.

The real advantage of these materials, however, is in their stiffness per unit weight. It is this characteristic that points toward their immediate applications, which are in the developmental stage. Among them are composites consisting of carbon fibers in epoxy resin for strong and stiff compressor blades in lightweight jet engines, and boron in epoxy resin for helicopter rotor blades that turn at high speed.

When it is not important to hold down weight, tungsten fibers, which retain high strength up to more than 1,500 degrees C., can be used in a composite. Tungsten wires have been introduced into metal matrices for use at temperatures of 1,000 degrees and higher. The metals cobalt and nickel make ideal matrices because they do not oxidize readily at high temperatures.

Fibers of tungsten, silica coated with carbon, graphite and boron are put into matrices by electroplating. Alternatively the matrix is chemically deposited on the fibers, a technique that avoids damaging them.

Another technique does away altogether with the problems inherent in making fibers and matrix separately and somehow joining them to form a useful composite. The technique makes the matrix and the reinforcing fiber in one operation by the controlled melting of certain metal alloys. In these "eutectics," which take their name from the Greek word for "easily melted," part of the alloy develops into parallel whiskers and part becomes a matrix for the whiskers. The result is a whisker-reinforced composite of great strength and good heat-resisting properties. For example, a eutectic developed by Frank Lemkey and Michael J. Salkind at the United Aircraft Research Laboratories consists of niobium carbide whiskers in a niobium matrix; the composite has demonstrated high strength at temperatures up to 1,650 degrees C.

The practical application of new composite materials is inextricably linked with problems of engineering design. The new materials are ideally suited for carrying large tensile loads in one direction but may not be as effective as other materials under compression or under shear occurring at an angle to the fibers.

Careful design can overcome such deficiencies. A case in point is a hollow glass sphere needed as a buoyancy tank in a deep-diving submarine. The structure would have been very difficult to make from a single piece of glass, and thus constructed it would have been extremely dangerous when not under pressure. A U.S. Navy research group made a suitable structure with sections of glass-reinforced plastic. In it the fibers run radially, so that they support one another against buckling when the vessel is under pressure.

The superior tensile strength of fiber composites makes them ideally suited for use in certain advanced engineering concepts. One such concept is stressed-skin construction, in which the skin of a structure is always in tension, even under compressive forces. (It may be helpful here to think of a football with someone sitting on it; the covering of the football is under tension from the pressure of the air inside it even when the football itself is under compression from the weight of the sitter.) Prestressed structures of the kind that have been made from concrete will also be able to make use of the new composites. As these materials come into general use, they will provide the basis for buildings with the contours of tents and aircraft with the delicate and graceful lines of birds.

How is heat conducted through a material? The key is the phonon, a particle-like packet of waves that can travel through a solid although the atoms in the crystal lattice are anchored in place.

JOHN ZIMAN

The thermal properties of materials

The management of heat flow is a major concern of the practical engineer. He wants heat to travel freely from flue gas into the boiler and from steam into the condenser. He tries hard to keep it out of the refrigerator. Thermally conducting and insulating materials are essential to the functioning of a modern industrial society. Merely to calculate the transfer of heat in a complex structure, given the basic thermal properties of the constituent materials, is a formidable task. For such purposes the engineer is usually content to work with tabulated data on materials of every kind, from asbestos, concrete and high-alloy steels to polyurethane foams and window glass.

Can a more analytical approach be achieved? What will be the effect on the thermal conductivity of the metal walls of the heat-exchanger in a nuclear power station if the alloy mix is subtly changed? What proportion of tellurium to selenium in their compounds with bismuth will make the best thermoelectric generator? What types of material are best suited, in principle, for devices to store heat produced from off-peak electric power? A vigorous science dealing with the thermal properties of materials of all sorts is obviously highly to be desired.

For an exact organized science we must build on a firm basis. We need, first, an understanding of the thermal properties of relatively perfect crystals of pure elements or of relatively simple chemical compounds. We must start, indeed, from an understanding of the nature of "heat" itself.

Isaac Newton and his contemporaries knew very well that heat is a mode of motion. But the quantitative experiments of Joseph Black in the middle of the 18th century, establishing that a given mass of matter has a specific capacity for taking up heat, were so persuasively in favor of heat as a fluid (caloric) that it took another century for the dynamical explanation to be reinstated. This explanation, embodied in the classical physics of James Joule, Hermann von Helmholtz, Lord Kelvin, James Clerk Maxwell and Ludwig Boltzmann, was very sound with respect to gases. With the realization that all the properties of gases were primarily kinetic—that they exerted pressure, could take up heat and so on because the gas molecules were all in motion—one had gone most of the way toward a quantitative theory.

Suppose, for example, that each molecule is an idealized structureless particle. One can then show by a general statistical argument that the average energy of each molecule is proportional to the absolute temperature (T) in the relation $3/2\,kT$, where k is the universal Boltzmann constant (1.381×10^{-16} erg per degree centigrade). This means that the heat required to raise the temperature of a volume of gas containing N such particles by one degree would be $3/2\,Nk$. Thus the specific heat, or heat capacity, of an ideal gas is the same for a given number of molecules, and is independent of temperature.

The conduction of heat by gases is also easy to calculate. Because the molecules are in rapid motion, they diffuse throughout any space they occupy. Heating one side of a box of gas increases the average energy (that is, velocity) of any molecules hitting that wall. These "hotter" molecules diffuse into other regions, where they share their excess energy with "cooler" molecules. Even though there may be no net flow of particles from one region to another, energy is transported from the hot wall to the cold wall of the box; we say that heat has flowed through the gas.

From this description one can see that the thermal conductivity of a gas must be proportional to its specific heat (C) and to the average speed of the molecules ($\bar{v}$). In addition, and most significantly, the rate of heat transfer will be proportional to the distance each molecule travels between collisions. The larger the value of this "mean free path" (L), the farther apart are the regions exchanging energy and therefore the larger is the energy difference to be shared in the final collision. Combining C, $\bar{v}$ and L with a geometrical factor of 1/3, one arrives at the standard kinetic formula for the thermal conductivity (κ) of a gas: $\kappa = 1/3\ C\bar{v}L$.

The really difficult calculation is finding the mean free path, which depends not only on the number of molecules per unit volume and their collision "cross section" but also on the relative motion of the colliding particles and on their range of speeds. Nevertheless, the above formula contains essentially all the physics of the situation.

Very well; now apply the same reasoning to heat transfer in a crystalline solid. The problem foxed classical physics. The specific heat was easy enough to calculate. Think of each atom as vibrating inside the cage of its neighbors, to which it is also connected by chemical bonds. For simplicity consider each atom as moving independently of its neighbors. Energy is stored both as kinetic energy in each vibrating atom and as potential energy in the bonds that are continuously being stretched or compressed as the atom vibrates. On this argument the specific heat of a solid is simply $3\ Nk$, or just twice the value for an ideal gas containing the same number of atoms. Nothing could be simpler, and this very result had been discovered experimentally by Pierre Dulong and Alexis Petit in 1819, more than 50 years before statistical mechanics was invented.

Unfortunately a quantitative theory of heat conduction in solids cannot be built on this basis. The model is much too simplified. By throwing away all the effects of the movement of an atom on its neighbors, one has destroyed the mechanism by which, in fact, heat energy passes from one oscillating atom to the next throughout the crystal.

With the development of statistical mechanics it became possible to link the thermal conductivity of metals with their electrical conductivity. In order for metals to conduct electricity readily it seemed clear that they must contain a dense gas of highly mobile charged particles, subsequently identified as electrons. Since each charged particle can carry heat energy in addition to its electric charge, it follows that the thermal conductivity of a metal should be proportional to its electrical conductivity multiplied by the absolute temperature. This relation, again, had been discovered experimentally by Gustav Wiedemann and Rudolph Franz in 1853.

Paradoxically, this free-electron theory of metals, as it became established around 1900, presented a grave difficulty for the theory of specific heat. We have seen that the specific heat of metals, as shown by Dulong and Petit, is $3k$ per atom. If the free electrons in a metal behaved as classical particles, each would make an extra contribution of $3/2\,k$, with the result that the specific heat of metals would far exceed the experimentally observed value.

This puzzle was solved only by the development of quantum mechanics after 1925, and by the recognition that a very dense electron gas satisfies laws of statistical mechanics quite different from those of an ordinary classical gas and as a result has a very small specific heat. In such a system the whole basis of calculating electrical conductivity has to be changed. It turns out, however, that the new basis of calculation yields the Wiedemann-Franz relation at room temperature and above. The theory of heat conduction in metals has thus become an offshoot of the theory of the electrical properties of matter [see "The Electrical Properties of Materials," by Henry Ehrenreich, page 127].

Any mechanism that can scatter electrons and give rise to electrical resistance contributes proportionately to thermal resistance. Indeed, in semiconductors, which are intermediate between metals and insulators, a significant fraction of heat is often carried by electrons or other mobile carriers. The amount of heat conducted in this way significantly limits the performance of thermoelectric generators (devices that convert heat directly into electricity) and thermoelectric refrigerators (devices that use electricity to drive heat directly from one region to another).

At ordinary temperatures all materials—metals as well as insulators—store heat as a vibratory motion of their atoms. What happens when the temperature is very low? As Albert Einstein pointed out in 1907, Max Planck's hypothesis that this vibrational energy, like all energy, must exist in distinct quanta should have dramatic consequences. As the temperature of a solid was lowered, its heat capacity would fall much more steeply than one

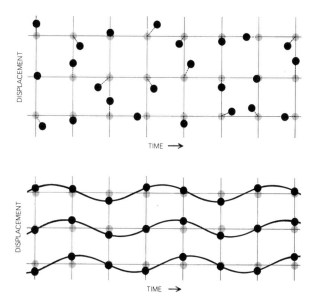

Heat conduction in solids differs from that in a gas because the atoms are tied to lattice positions. In 1907 Albert Einstein devised a formula for heat conduction based on the assumption that atoms vibrate independently (top). In 1912 Peter J. W. Debye argued that neighboring atoms, being bonded together, tend to vibrate in unison (bottom).

would expect from the classical relation $3NkT$. In fact, Einstein predicted that the heat capacity would become vanishingly small well before the temperature itself reached zero.

Although Einstein's analysis was substantially correct, careful measurements showed that the actual decline in specific heat was not as rapid as he had predicted. The explanation of this discrepancy between theory and experiment by Peter J. W. Debye, and almost simultaneously by Max Born and Theodore von Kármán, in 1912 was the key to the entire problem of the thermal properties of solids.

Einstein had assumed, for the sake of simplicity, that each atom in a crystal oscillated independently. Debye saw that one must allow for the coupling forces that tend to make neighboring atoms move together. This seems to present a fantastically difficult mathematical problem. But think of the crystalline solid as a continous medium, without any fine-grained atomic structure. One knows that sound waves of various wavelengths can be excited in any such solid, and that their velocity will depend on bulk properties such as density and elasticity. Why not, then, describe the heat motion of atoms in terms of a medley of elastic waves batting around inside the solid?

This is what Debye proceeded to do. He treated each different mode of vibration—corresponding to waves traveling in a different direction and with a different polarization or a different wavelength—as an independent dynamical system, to which the basic rules of Planck's quantum theory could still be applied. Now, however, we have many different frequencies right down to acoustic vibrations of the crystal. As a result the low-temperature cutoff in specific heat is somewhat abated by the existence of low-frequency modes of vibration (whose energy quanta are very small) in which heat can still be stored. This led Debye to a formula for specific heat in which the value falls as the cube of the absolute temperature. The formula is known as the T^3 law of specific heats, and it was one of the early triumphs of quantum theory.

The Born-von Kármán theory was a more rigorous version of the ideas used by Debye. In many ways their analysis is a complete formal solution of the problem of the specific heat of solids. Given the forces between atoms, one can calculate the spectrum of vibrational frequencies and then compute the specific heat as a function of temperature. (The fact that a crystal system is translationally invariant, meaning that each repeating unit cell of atoms is a replica of its neighbors, allows one to use a theorem of Felix Bloch's that reduces the mathematics to a problem involving only one cell at a time.) By counting modes of vibration one automatically obtains the Dulong-Petit law at high temperatures. At the same time the Born-von Kármán equations contain the elastic-wave solutions required for Debye's T^3 law at low temperatures.

For some years the Born-von Kármán equations provided the basis for a vigorous cottage industry: one tried to validate assumptions about the interatomic forces in solids by proving that they gave rise to the observed variation of specific heat with temperature. This laborious and elastic chain of argument has largely been superseded by direct methods of observing individual vibrational modes, such as the inelastic diffraction of neutrons.

As Debye himself showed in 1914, these early quantum theories resolve the major difficulty in the calculation of the thermal conductivity of solids. Each vibration of a crystal lattice can be described as a traveling wave carrying energy. By analogy with the

Thermal properties of materials 117

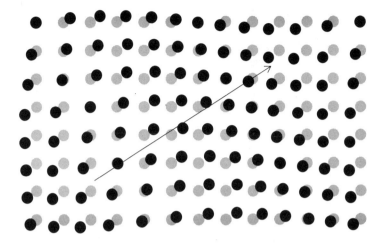

Lattice waves in crystal, represented by alternate bunching and separation of atoms, can be inferred from Debye's hypothesis that adjacent atoms in a crystal influence each other's motion. According to this model, thermal energy should travel through a crystal in waves similar to those that carry acoustic, or mechanical, energy. Thermal waves have a higher frequency, however, than any that can be produced mechanically. In accordance with quantum theory such lattice waves can be treated as particles, or phonons.

quantum theory of light, according to which light waves exhibit the attributes of particles when they are treated as photons, these traveling sound waves can be treated as particles called phonons. This single quantum principle applied to an otherwise purely classical argument allows us to think of the crystal as if it were an empty box through which phonons travel virtually without hindrance, carrying heat as the molecules of a gas do.

Indeed, the conduction of heat would now seem to be altogether too easy. What would prevent the phonons from traveling from one side of the crystal to the other without let or hindrance? In a perfect crystal the mean free path of a phonon ought to be as large as the crystal itself, in which case heat ought to travel through a crystal as direct radiation, so to speak, from a heated surface to a cooler one.

But as Debye then argued, even ordinary sound waves do not travel through the most perfect crystal without being scattered, except at very low temperatures. The solid is in ceaseless thermal fluctuation, so that at any given moment there will be regions of higher and lower density distributed throughout its volume. The velocity of sound, however, in an elastic medium depends on density. The heat-carrying phonon finds itself being propagated through a medium whose properties vary irregularly from point to point; therefore it is deflected and scattered. The phonon's mean free path is thus limited, and can be calculated roughly.

It is easy to show that the rate of scattering of a phonon ought to be proportional to the mean square of the various fluctuations in density, and that this value depends directly on the absolute temperature. To calculate the constant of proportionality, we need to know how much the velocity of sound varies with volume when a solid is compressed or when it expands. In a theory of thermal expansion published in 1912, E. Gruneisen had shown that if there were such an effect, say of magnitude gamma, then the coefficient of thermal expansion ought to be just gamma times the compressibility multiplied by the specific heat. The general idea is that, when heat is fed into a crystal, it pays for the crystal to expand a little so as to reduce the size of the quanta in which the vibrations of the crystal lattice store energy. To calculate gamma itself one would have to know exactly how the interatomic forces vary with distance, which is quite a subtle problem. For most ordinary solids, however, the actual coefficient of expansion leads to a value of about two for gamma.

Another achievement of the Einstein-Debye theory was a general formula for the melting temperature of solids. This formula was proposed by Frederick Lindemann (later Lord Cherwell), who suggested that a solid must melt when the average amplitude of vibration of each atom reaches some definite fraction (empirically about a tenth) of the diameter of the unit cell within which it is caged. One can calculate the melting point in terms of the velocity of sound in the solid (or some other measure of the elastic forces on the atom) and the atomic volume. Again the result is in surprisingly good agreement with experiment for a wide range of substances.

Looking back on this period just before World War I we see that a quantitative theory of the thermal properties was already firmly established. It had been achieved by adding only a few simple quantum ideas to classical mechanics. At ordinary temperatures the thermal conductivity of a nonmetallic crystal ought to be given by the kinetic formula, that is, the gas of phonons may be supposed to have the specific heat $3\,Nk$ and the phonons themselves an average velocity equal to the velocity of sound in the solid. The Debye, Gruneisen and Lindemann theories can then be combined to yield a rather simple formula for the phonon mean free path (L), which ultimately determines the rate of heat flow. After making allowances for complicated geometrical scattering, one finds that L is approximately equal to $20\,T_m d/\gamma^2 T$, where T_m is the melting point, T is the absolute temperature, γ is Gruneisen's gamma and d is the dimension of the crystal lattice. This formula demonstrates the well-known fact that thermal conductivity falls as the temperature rises. The formula also agrees with the observation that at room temperature the thermal conductivity of diamond—because it has a high melting point and permits a high phonon velocity—is comparable to that of metallic copper. For all its apparent naiveté, the formula agrees as well with experiment as anything physicists have since been able to derive by more sophisticated and vastly more complicated arguments. Yet we now know that it has grave conceptual weaknesses.

The basic objection, put forward by Rudolf Peierls in 1929, is that the "thermal" fluctuations of density that scatter the heat-carrying phonons are themselves phonon models. We must then ask the question: What happens when two phonons meet? In an idealized "harmonic" crystal, where by definition the lattice modes are dynamically independent, there is no interaction. But in any real solid (where, for example, Gruneisen's gamma is not zero) the interatomic forces are not harmonic, and two traveling waves can interfere with each other. When this happens, the two waves can combine to produce a wave whose frequency is the sum of their individual frequencies. In quantum language we say that two phonons have collided and been destroyed, and that in their place a new phonon has been created that conserves the energy and momentum of the original particles [*see illustration on page 120*]. Conversely, a single phonon can spontaneously split up into two new ones, again conserving energy and momentum.

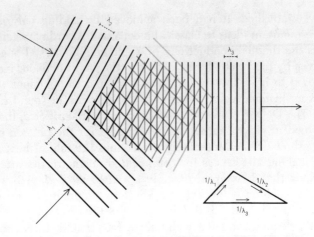

Interaction of two phonons gives rise to a new one whose frequency is the sum of the frequencies of the two interacting phonons. This is the normal, or "N," process, in which momentum is conserved. The reaction is represented by the vector diagram at the lower right; the sides of the triangle correspond to frequency. It is also possible for a phonon to split up into two new ones, with energy and momentum again being conserved.

At first this would seem to be just what we need to make our "phonon gas" model of heat conduction even more realistic. The collisions that limit the mean free path of molecules in a real gas are mimicked by the interactions of phonons. The analogy, however, is not exact. When molecules collide, the excess energy of one of them is shared, so that eventually, after several collisions, the energy is dissipated as an increase in the energy of other molecules moving in random directions. Colliding phonons have the unfortunate property of always handing on their energy and momentum to their successors in such a way that the net current of heat is neither decreased nor deflected by the process. If all phonon-phonon interactions were of this kind, called normal processes, the thermal conductivity of a crystal would be infinite.

Peierls resolved the paradox by showing that the interference of traveling waves in a crystal lattice is not the same as it is in a continuous medium. Suppose we have two waves of short wavelength (high frequency) traveling in the same direction. Their combination wave ought to have an even shorter wavelength (higher frequency) and also ought to be moving in the same direction. But if the new wavelength is shorter than twice the lattice spacing, an ambiguity arises. The motion of the atoms in the lattice no longer tells us clearly which way the new wave is traveling. In fact, the motion of the atoms is entirely consistent with a wave of substantially longer length moving in the *opposite* direction [*see the illustration on page 122*]. In other words, the "momentum" ascribed to the phonon is only a Pickwickian "crystal momentum" and need not be conserved in phonon-phonon interactions.

To describe such reversals in crystal momentum Peierls applied the term *Umklapp*, which in German means "flopover." Thus physicists distinguish *Umklapp* (*U* processes) from normal processes (*N* processes) in phonon-phonon interactions. The *U* processes are quite effective in reducing the mean free path of phonons. It can be shown that *U* processes occur at a rate proportional both to the absolute temperature and to the square of the strength of the nonharmonic forces, as characterized by Gruneisen's gamma. In principle, therefore, Debye's formula for the mean free path remains valid, modified only by the *Umklapp* concept.

As so often happens in theoretical physics, the gain in conceptual understanding was not rewarded by greater ease of calculation. Except in crude approximations, which are not much better than one can obtain with the simple Debye formula, the new computations have not been carried through. Accordingly many finer points governing the differences between various classes of materials are not at all well understood.

Nonetheless, one prediction of the Peierls theory has been fully confirmed by experiment. As we have seen, the thermal excitation of high-frequency lattice modes is difficult at low temperatures. It is only these modes, however, that are short enough to flop over. Cooling a crystal therefore freezes out the *U* processes. As a result the mean free path of the phonons, and with it the thermal conductivity, rises dramatically. In fact, at liquid-helium temperatures heat conduction is effected by the "radiation" of phonons.

The effect does not have much practical significance in ordinary applications of materials science. Nevertheless, in 1956, while investigating thermal conductivity at low temperatures, Robert Berman of the University of Oxford was led to consider an impor-

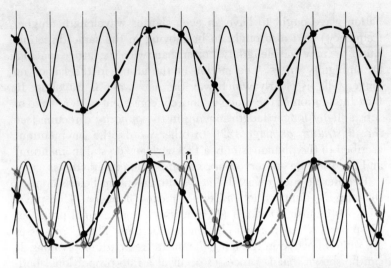

Ambiguity in wavelength arises when the vibrations in a lattice can be assigned a wavelength that is either shorter than twice the lattice spacing (black curve, top) or longer (dashed curve). The motion of the short wave to the right (bottom) is equivalent to the motion of the corresponding long wave to the left. Thus two high-frequency phonons can interact to produce a phonon that travels opposite to the expected direction. This "Umklapp," or flopover, process appears to violate conservation of momentum.

tant scattering mechanism that had been largely overlooked. The Peierls theory predicted that conductivity should increase exponentially as the crystal specimen was cooled. Berman found that some substances, such as solid helium and artificial sapphire (fused alumina), followed the expected law. In other substances, such as silicon and germanium, the increase of conductivity was much slower. It occurred to Berman that the deviation from theory was largest in materials with substantial concentrations of different isotopes. In potassium chloride, for example, the chlorine isotopes of masses 35 and 37 occur naturally in the ratio of three to one. Could isotope ratio be having an important effect on thermal conductivity? Berman and others soon demonstrated that such was the case.

An isotope substituted as an "impurity" in an otherwise perfect crystal must scatter phonons as though it were a little local mass

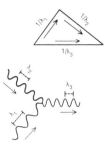

Unimpeded heat flow would result if all interactions between phonons were normal, or N, processes in which crystal momentum is conserved. Phonons created at the hot end of a perfect metal crystal (left) would travel swiftly to the cool end. There would be no thermal resistance. The interaction of two phonons in an N process is illustrated at the right together with a vector diagram.

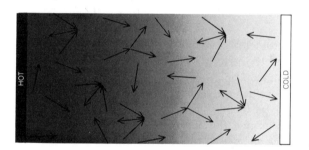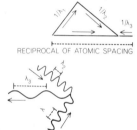

Actual heat flow in a metal crystal (left) is slowed by Umklapp, or "U," processes, which do not conserve crystal momentum and largely account for the thermal resistivity observed even in pure crystals of metal. A U process is represented at the right.

in an elastic medium. The scattering is proportional to the square of the difference in isotope mass and (like the scattering of light by air molecules that makes the sky blue) to the inverse fourth power of the phonon wavelength. Thus the longer the wavelength, the less the scattering. In fact, the long-wave phonons are so weakly scattered by isotopes that they seem able to transport all the heat without serious dissipation. To get a proper result one must invoke the phonon-phonon N processes, which do not directly give rise to thermal resistivity but which force the long-wave phonons to combine with one another, thereby transferring the heat energy into short-wave modes that *can* be scattered. In 1959 Fred Sheard and I, then working at the University of Cambridge (and, independently by a different method, Joseph Callaway, then working at the Westinghouse Research Laboratories), were able to derive formulas that matched up with the experimental results.

Actually isotopes are not an important source of thermal resistance in practical materials at ordinary temperatures, but the difficulties experienced in deriving quantitative theories of this very simple effect indicate the size of the problem of calculating the thermal conductivity of ordinary solids. We seldom have to deal with pure and perfect single crystals. What should we allow for the effect on phonons of chemical impurities, vacancies, interstitial atoms, grain boundaries, dislocations, stacking faults, magnetic-domain walls and all the other blemishes that real materials inevitably contain?

In addition to the problems of defining the physical situation, evaluating variables and trying to compute answers, the theory of the thermal properties of materials presents some difficult questions of general principle. Consider the scattering of phonons by a dislocation in an otherwise perfect crystal. It is easy to argue that the elastic strain produced around the dislocation creates changes of local density—and hence changes of the effective velocity of sound—that interact with lattice waves by the same mechanism that operates in the Debye theory of thermal conductivity. It has been found, however, that this calculation makes a much smaller contribution to thermal resistivity than is actually observed (even allowing for the fact that it is difficult to make accurate measurements). Most of the scattering, it appears, is the scattering of short-wave phonons at the core of the dislocation, where the arrangement of the atoms is very far indeed from the perfect lattice. The mere change of density is not sufficient to account for this; there must be a diffraction effect due to the actual differences in the relative positions of the atoms in this special neighborhood. Somehow or other short-wave phonons are so affronted by the unorthodox patterns of atoms at the core that they are strongly scattered.

An extreme case of the same general problem is presented by a truly amorphous material such as glass. What effect does the disordered arrangement of atoms throughout the material have on the propagation of elastic vibrations? I believe that a new mathematical technique invented by one of my colleagues, James Morgan of the Zenith Radio Research Corporation of Great Britain, may provide an entry into this subtle and hitherto intractable problem. This whole field involves classical mechanics, probability theory and a large helping of three-dimensional geometry.

In spite of these formidable mathematical difficulties, the basic physics of heat conduction is well understood. Physicists have complete confidence that it is governed by the same principle of quantum theory that rules all other solid-state phenomena. In one or two fields of materials science, such as the choice of materials for thermoelectric devices, theory has helped to guide the applied scientist and engineer. But it must be admitted that theory has a long way to go before it can design new materials to satisfy the demands of engineers employed in heavy industry. This lack of direct applicability does not detract from the study of this large and difficult topic for its own sake: for the knowledge that is acquired and the intangible rewards of traveling along a stony road.

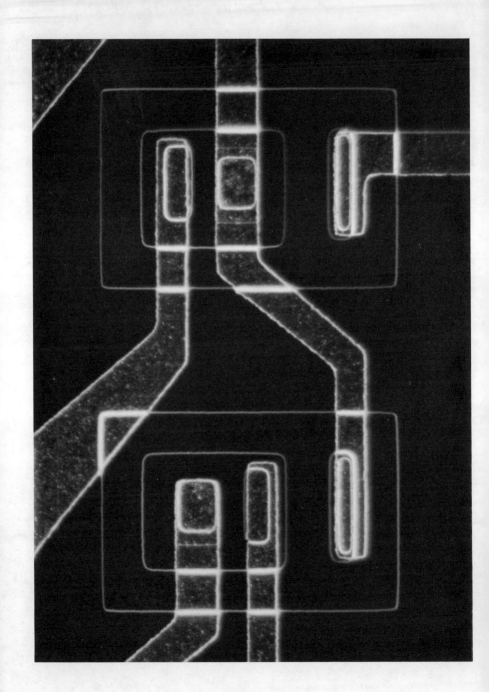

Materials differ in their resistivity to an electric current by as much as 23 orders of magnitude. The insights of quantum mechanics are helping to make this full range more accessible to technology.

HENRY EHRENREICH

The electrical properties of materials

The electrical conductivity of materials was first demonstrated in 1729 by an English experimenter, Stephen Gray, who touched a charged glass rod to the end of a moistened cord and discovered that the cord transmitted the electricity a distance of about 1,000 feet. Today the exploration of the electrical properties of solids is disclosing much more dramatic phenomena, and these have turned out to be of fundamental significance in the understanding of matter, as well as of great technological importance.

To begin with, the wide range of electrical conductivities exhibited by materials is itself a striking fact. Between the most

The two transistors in this photomicrograph (about × 900) are part of an integrated electrical circuit that contains 14 such transistors plus other components on a square clip of silicon about 1/20 inch on a side. The black areas are layers of silicon "doped" with impurities to produce regions that conduct by means of electrons and regions that conduct by means of electron "holes." The slightly lighter areas are aluminum that has been vapor-deposited over the entire surface of the silicon wafer and then selectively photo-etched to achieve the desired pattern. The circuit is manufactured by Fairchild Semiconductor.

conductive substances (copper and silver) and the most resistive (polystyrene, for example) the difference amounts to 23 orders of magnitude. To appreciate the extent of this spread, compare it with extremes in the scales of distance. One might note, for instance, that the ruler needed to measure the size of the universe is only some 23 orders of magnitude larger than the mile ruler that measures distances on the earth. Evidently, then, the electrical conductivity (or rather its inverse, resistivity, the quantity customarily used in statements of Ohm's law) is one of the most varying of all physical quantities.

The individual materials themselves show great variability in resistivity according to the conditions of temperature, pressure and the mixture of component substances. The addition of a minute trace of gallium or arsenic (one part per billion) to pure germanium increases its conductivity by two orders of magnitude (nearly 1,000-fold) and makes it suitable for use in transistors. A tiny further addition of the impurity can increase the conductivity 100,000-fold, converting germanium to a conductor of tunnel-diode grade. Similarly, silicon and metal oxides such as nickel oxide and titanium dioxide are lowered in resistivity by the introduction of appropriate impurities. Indeed, nickel oxide, which is an insulator in the pure state, is reduced in resistivity by 13 orders of magnitude by adding 1 percent of lithium.

Dramatic changes are also produced by changes in temperature. A semiconductor can be made a conductor by heating it to a high temperature or it can be made an insulator by cooling it to a low temperature. In contrast, the resistivity of a pure metal is much less drastically increased by heating and reduced by cooling. In some instances the change is very abrupt. For example, above 150 degrees Kelvin vanadium sesquioxide is a semicondutor; when it is cooled to just below that temperature, its resistivity suddenly jumps by six orders of magnitude and it becomes a good insulator! The best-known kind of sudden transition, of course, is the total disappearance of resistivity in certain metals (that is, their sudden transformation to the superconducting state) when they are cooled to temperature near absolute zero.

The impact of light, like changes in temperature, also can affect electrical resistivity. Some semiconductors and insulators are exeremely sensitive to light, and under illumination their conductivity may be several orders of magnitude higher than it is in the dark. This phenomenon is called photoconductivity.

How are these facts to be explained? What principles account for the great differences in conductivity between metals and insulators, the peculiar properties of semiconductors, the abrupt transitions, the potent effects of impurities, temperature and light?

The first reasonable approach to an explanation of electrical conduction in solids was proposed in 1900 by the German physicist Paul Karl Drude. His crucial contribution was the recognition that current passing through metals must be carried by the charged particles (now called electrons) that J. J. Thomson had discovered just a few years earlier. Drude imagined conductive metals to be permeated by a gas of free electrons. An applied electric field, he suggested, accelerates the electrons along the field; because the ions they encounter in the crystal lattice deflect them and thus interpose resistance, the electrons settle into a constant drift velocity that is proportional to the strength of the applied field. (The situation would be similar to the fall of an object from an airplane: the object falls with accelerating speed until the gravitational force is balanced by the force of friction with the air; thereafter the fall continues at a constant velocity.) Drude supposed the mean free path of the electrons between collisions to be only about the length of the distance between atoms in the crystal; to explain the observed conductivity of metals he assumed that all the electrons were free to act as carriers.

Drude's model successfully accounted for several observations, including the fact that in many metals the transport of heat is proportional to the transport of electricity at a given temperature, which could now be explained by assuming that electrons are involved in the transmission of heat. In certain predictions, however, the Drude hypothesis proved to be incorrect. It failed to account for the observed variations of conductivity with temperature, and since it required that all the electrons must be free, it implied that the electron gas should have a higher specific heat than was actually found to be the case; experiments showed that the amount of energy needed to raise the temperature was so small that one had to conclude only a very small fraction of the valence electrons, rather than the entire gas, was involved in conduction.

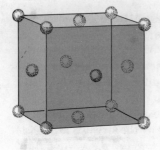

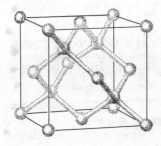

Intrinsic charge carriers, typically valence electrons (electrons from the atoms' outermost electron shell), can conduct an electric current in pure crystals. In a metal that has one valence electron, such as copper (top), the valence electrons (gray) are spread out over the entire lattice of the crystal. It is the availability of these "delocalized" electrons that makes such metals good electrical conductors. In a semiconductor, such as germanium (bottom), the four valence electrons cement the atoms of the crystal together in a diamond-like structure by covalent bonds. At absolute zero these "localized" electrons are not free to act as electrical carriers and germanium is an insulator. In order for pure germanium to become conducting, some of its chemical bonds must be broken and the bound valence electrons released; this can be done by supplying energy to the crystal in the form of heat or light.

Drude's concept of mobile electrons as the agents of conduction in a metal was essentially correct, but many of the details of his model were not. The answers to the difficult questions it raised had to wait for the advent of quantum mechanics a quarter-century later. Some of the quantum-mechanical considerations affecting electrical conductivity are discussed by Sir Nevill Mott in the second chapter of this book, "The Solid State." Let us look into the situation in further detail.

The first requirement for conduction is a supply of carriers that are free to wander through the solid. In certain solids (such as ice and silver bromide) electricity is transmitted by ions, but in most cases, and particularly in metals, the carriers are valence electrons —electrons of the atoms' outer electronic shell. Taking a chemical view of the situation, one might describe the difference between conductors and nonconductors in terms of the relative availability of carriers. Consider the simple case of copper, a metal with a single valence electron. In the isolated atom the electron is spread out in a cloudlike orbit around the nucleus. In a crystal of copper, in which the atoms are tightly packed together, the electrons spread themselves over the entire lattice [*see illustrations on facing page*]. They find this energetically favorable because, according to the uncertainty principle, their delocalization lowers their kinetic energy. It is this effect that causes the atoms in the crystal to stick together. Such delocalized electrons are ready candidates for acceleration in an electric field.

In contrast to copper, the atoms of the semiconductor germanium are cemented together more favorably by forming covalent bonds. In the resulting diamond-like structure the electrons are not free to wander through the crystal or act as electrical carriers. Accordingly at absolute zero germanium would be an insulator. If, however, sufficient energy (in the form of heat or light) is supplied to break some of the chemical bonds and so release electrons, germanium becomes a conductor. Besides the electrons, the "holes" they leave at the vacated sites also become mobile electrical carriers, and under the influence of an electric field these move, like positive particles, in the direction opposite to that of the electrons. With increasing temperature the number of carriers in a semiconductor, and hence the conductivity, grows exponentially (whereas the conductivity of a metal such as copper, which has a large supply of carriers at all temperatures, is much less influenced by heat). Thus the bond-breaking requirement helps to explain why a semiconductor's conductivity increases so dramatically when it is heated.

The chemical view also accounts in a general way for the ability of a minuscule amount of impurity to transform a semiconductor into a much better conductor. The impurity simply adds readily available electrons to serve as carriers. A good example is provided by the "doping" of germanium with arsenic. The arsenic atom has five valence electrons to germanium's four. The fifth electron does not enter into the covalent bonds with the germanium atoms in the crystal, and therefore its bond to the parent arsenic atom is easily broken. Since little input of energy is needed to release these electrons, the doped semiconductor becomes conducting at much lower temperatures than pure germanium. And since only a comparatively small number of carriers is required in a conductor, a minute

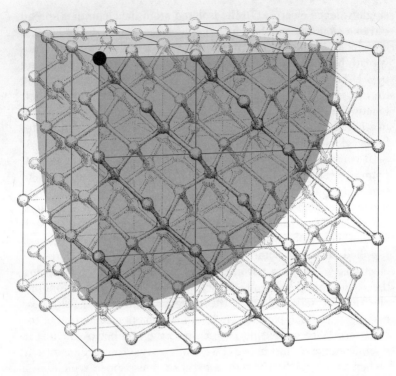

Extrinsic charge carriers can be supplied to germanium by doping the crystal with a small amount of arsenic. The arsenic atom (black) has five valence electrons to germanium's four. The fifth electron is only weakly bound to the arsenic atom and ranges over considerable distances from it, as indicated by the large gray volume. Only a small amount of energy is required to release the extra electron; as a result the doped crystal becomes conducting at much lower temperatures than pure germanium.

injection of arsenic can raise germanium's conductivity to nearly that of a metal.

In a general way, then, the consideration of the availability of carriers furnishes a useful picture of the distinction between conductors and nonconductors. In the case of metals a very small electric field can supply enough energy to free electrons so that they can drift downstream under the influence of the field. For a semiconductor a substantial amount of energy must be applied to break the chemical bonds and release electrons. Most frequently the energy supplied is nonelectrical. If it is electrical, the crystal breaks down as a condenser does when it is subjected to too large a voltage. An insulator differs from a semiconductor simply in the strength of the chemical bonds; in insulators the bonding of electrons is so firm that they cannot be freed by heat short of temperatures that will melt or evaporate the crystal.

How can one account for the transition from the insulating to the conducting state in crystals? Is the transition generally gradual (perhaps through a progressive merging of the lowest energy state of the crystal and the higher energy states) or is it abrupt? In vanadium sesquioxide we see such a transition occurring abruptly. Perhaps in such cases an abrupt change in the crystal's mechanical or magnetic structure is responsible for the transition. Can this be taken as the general rule for all transitions? The answer is not yet known for certain. Obviously the question of the mechanism of transitions is of great theoretical and technological importance.

Such general ideas, interpreted in the light of quantum theories concerning the excitation of electrons, are helpful in gaining insight into some of the basic questions about the electrical properties of materials. They are only an entering wedge, however, into the major underlying question of how one describes the energy of a crystal in terms of the quantum states of its electrons. Because of the large number of electrons in a solid, this presents a many-body problem of a most difficult kind on whose understanding electrical transport theories depend. The calculations undertaken to explore the conduction process have had to depend on models that represent approximations.

The most useful approximation that has been developed to describe the distinction between the conducting and the nonconducting states of a crystal is the band model. It is based fundamentally on consideration of the quantum states allowed to electrons by the Pauli exclusion principle. According to this principle, a given quantum state cannot be occupied by more than two electrons (with opposite spins). The principle accounts for the shell structure of isolated atoms. Since electrons prefer the lowest available energy level, each successive shell is filled before higher ones are occupied.

When atoms are grouped together in a crystal, there is an analogous situation. The electrons wandering through the crystal have energies that fall within "bands" derived from atomic shells. Within a band the difference between permissible levels is so infinitesimal that an electron can very easily be excited from one level to the next. The bands are separated, however, by gaps that are forbidden to the electrons [*see illustration on the facing page*].

In a metal the uppermost band, called the conduction band, is only partly filled. Hence an applied voltage can excite some of the electrons to heightened energy, so that instead of drifting at random they will move in the direction of the field. In an insulator or pure semiconductor, on the other hand, the conduction band is empty, and all bands below it are completely filled. To promote an electron across the gap to the conduction band from the uppermost filled band (called the valence band) requires a great deal of energy, which must be supplied by some external source, for example the input of heat or light. (Even an insulator, with its very large energy gap, can become conducting in the presence of light. Indeed, copying by xerography depends crucially on this process in one of the best insulators, amorphous selenium.)

This picture accounts for the presence of excitable and mobile electrons in a conductor, but we still have to explain the ability of the electrons to travel along the field against the "frictional" resistance that Drude attributed (incorrectly, as it turns out) to the stationary ions in the crystal lattice. The Drude model went wrong in its two principal assumptions; first, that the stationary ions can scatter the conduction electrons, and that therefore the mean free path of such electrons in a metal is only the distance between atoms, and second, that all the electrons are involved in the conduction process. The quantum-mechanical account of conductivity explains why both postulates are incorrect.

Why is the electrons' mean free path much longer than Drude supposed? The answer lies in the fact that electrons, like other particles, partake of the nature of waves. According to quantum mechanics, an electron traveling in a crystal can be described as a wave that is modulated by the electrostatic influence of the ions it encounters. As the wave comes near an ion, it is distorted into rapid wiggles that imply the electron is accelerated to a higher kinetic energy during its passage [*see illustration on page 136*]. This

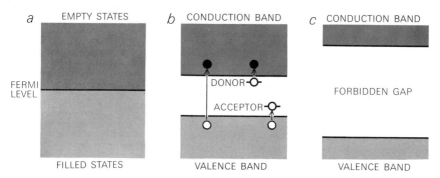

Energy-band structures of a metal (a), *a semiconductor* (b) *and an insulator* (c) *are compared in these diagrams. The electrons in all three types of solid can exist only in certain "allowed" energy bands, which are separated by "forbidden" energy gaps. The conduction band of a metal is partially filled with electrons even at a temperature of absolute zero. The highest energy occupied by the electrons is called the Fermi energy* (broken line). *In a pure semiconductor at absolute zero the valence band is completely full and the condution band completely empty. An input of energy to the semiconductor in the form of light or heat can raise a valence electron into the conduction band. Carriers can also be supplied to the semiconductor by impurity atoms with energy levels in the forbidden gap; "donor" impurities contribute electrons to the conduction band, whereas "acceptor" impurities contribute electron "holes" to the valence band. The forbidden gap in an insulator is too large to be bridged thermally short of melting the crystal.*

means that the electron spends comparatively little time near the ionic cores in the lattice and therefore is not greatly influenced by them. Indeed, in many simple solids the conduction electrons can be described as free particles, as Drude suggested, but with an apparent mass that may differ from that of electrons in a vaccum.

There is a crucial difference, however, between the classical picture and the one described by quantum mechanics. Whereas the Drude model implied that the mean free path of the electron between collisions should be of the order of the distance between atoms, the quantum-mechanical picture implies that the electron wave function adjusts itself to the cores in a systematic fashion. Consequently an electron can travel through a perfect crystal without ever being scattered! In other words, quantum mechanics predicts that in an ideal, fixed lattice the ions interpose no fric-

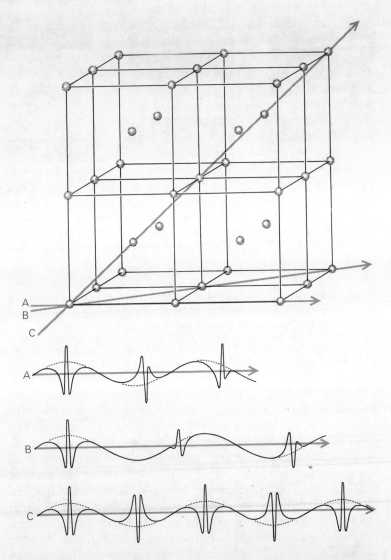

Wave function of an electron moving in a crystal lattice is modulated by the electrostatic influence of the ions it encounters. In this illustration the three different wave functions (solid curves below) correspond to three different directions through a body-centered-cubic lattice of sodium (diagram above). In all three cases the total energy of the electron (kinetic energy plus potential energy), and hence its wavelength (broken curves below), are the same.

tional force against the electrons' travel and the electrical conductivity is infinite. Of course, real solids are never perfect in this sense. Even if a crystal were completely free of impurities and defects such as dislocations, the ordinary thermal vibrations of its atoms would still cause enough irregularity to scatter the electrons to some extent. Nevertheless, the particles' wave characteristics indicate that the mean free path of the electrons must be considerably longer than the distance between atoms. (Indeed, electrons in highly purified gallium at very low temperatures have mean free paths up to 100 million times larger than the interatomic distance.) This larger mean free path in turn implies that the electrons have sufficient mobility to account for the observed conductivity of metals even though the number of contributing carriers is relatively small.

Why is it that only a small fraction of the available conduction electrons are involved in conductivity? Quantum mechanics again provides the answer. We start with the Pauli exclusion principle, which states in this context that only two electrons (of opposite spin) can have the same position in a space whose axes are the velocity components of the electrons' motion. (Velocity is defined as speed plus direction.) There is therefore a wide distribution of of velocities for the conduction electrons, with speeds ranging from zero to a certain maximum. The electrons with the maximum velocity form a boundary, called the Fermi surface, that marks the division between occupied and unoccupied regions of velocity space [see "The Fermi Surface of Metals," by A. R. Mackintosh; SCIENTIFIC AMERICAN, July, 1963].

In the stable electronic configuration of the crystal there are as many electrons traveling with a given speed in one direction as there are traveling with the same speed in the opposite direction; therefore there is no net velocity for the system as a whole. If we

The rapid wiggles of the waves in the vicinity of an ion imply that the electron has a higher kinetic energy there than elsewhere. This in turn means that the electron travels very rapidly through this region and spends comparatively little time in it. As a result the influence of the ionic cores is not nearly as large as one might expect, and in many simple solids the conduction electrons can be described as if they were free particles.

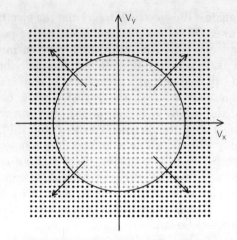

Fermi surface in a metal marks the division between occupied energy states (gray dots) in "velocity space" and unoccupied states (black dots). In this simplified two-dimensional diagram of a Fermi sphere each dot corresponds to an allowed electron velocity with components V_x and V_y. According to the Pauli exclusion principle only two electrons of opposite spin can have the same velocity. The radial arrows represent the direction of increasing speed and therefore increasing energy. Conduction electrons in the solid are accommodated at the lower energies first. In the absence of an electric field there is no net velocity for the system as a whole.

now apply an electric field, it will accelerate the electrons (in the opposite direction because of their negative charge). In a short time each electron will have added to its velocity vector a small increment parallel to the field. This change overturns the previous balance of velocities and can be pictured as a shift of the Fermi sphere [*see illustration on facing page*]. If there were no resistance in the crystal, the Fermi sphere would continue to shift, with more and more velocities in the downfield direction and fewer and fewer upfield. However, the thermal motion of the ions and impurities that scatter the electrons from their downfield paths tend to turn some of them to velocities headed back upfield. When this resistance effect matches the accelerating effect of the applied electric field, the Fermi sphere ceases to shift. The net shift that has occurred when this steady state is reached measures the drift

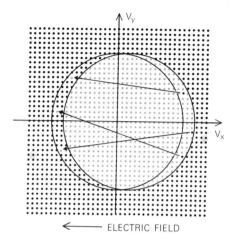

Fermi surface shifts under the influence of an applied electric field. Electrons in the metal drift preferentially in the opposite direction from that of the field. The net effect is that some electrons are transferred from the back of the Fermi surface (corresponding to velocities pointing to the left) to the front of the Fermi surface (corresponding to velocities pointing to the right). Collisions of the electrons with impurities and with atoms displaced by lattice vibrations (arrows) oppose this tendency, stabilizing the downfield shift of the Fermi sphere. Since all the electrons at the surface have the same energy, these transfers entail little cost in energy.

velocity of the gas of electrons in the downfield direction. The drift velocity can be increased by applying a stronger field or by lengthening the electrons' mean free path.

The scattering of the electrons only changes their energy by a small amount. Such a scattering will take the electron from an occupied to an unoccupied velocity state, however, and since the drift velocity is far smaller than the speed of electrons at the Fermi surface, only those few electrons very near the Fermi surface can be scattered. In fact, one can look at the entire process as if only a small fraction of all the electrons were participating in the conduction process. A similar argument shows why the electron contribution to the specific heat of a metal is as small as is observed.

With these facts in mind, let us consider some of the important scattering mechanisms in the lattice (involving heat and imperfections) and their influence on the electrons' mobility at various temperatures. I have already mentioned the thermal vibrations of the lattice atoms in this connection. At moderate temperatures the amplitude of their oscillations around their equilibrium positions is small enough to allow the electrons a mean free path considerably longer than the distance between atoms. As the temperature rises and the oscillations become more violent the mean free path is shortened and the conductivity of the crystal of course is reduced. Conversely, a lowering of the temperature increases the mean free path and the conductivity. There are limits, however, to the conductivity that can be achieved by cooling (except in the special case of superconducting metals at very low temperatures); these limits are imposed by imperfections, which are present in all real crystals.

The most common imperfections in many semiconductors and metals are impurities (foreign atoms). Their scattering effect as a function of temperature is rather different in semiconductors from what it is in metals. In metals (where the conduction electrons have relatively high velocities at all temperatures, since their velocity is determined primarily by the Fermi surface rather than by temperature) the scattering effect of impurities is constant, regardless of temperature. In contrast, in a semiconductor the effect of impurities varies considerably with temperature. At low temperatures an electron moves past an impurity slowly, spends a comparatively long time in its vicinity and therefore is scattered more effectively. At higher temperatures, with the electrons at correspondingly increased velocities, the impurities become less effective scatterers. Hence we expect the mean free path of a semiconductor to increase with rising temperature until the lattice vibrations due to the high temperature become so violent that they have an important scattering effect. In metals, on the other hand, the mean free path is constant up to temperatures at which the scattering effect of the lattice vibrations becomes dominant.

At extremely low temperatures, oddly enough, lattice vibrations can play quite an opposite role. Paradoxically, vibrations of the lattice are responsible for the infinite conductivity of superconductors. An electron traveling through the lattice at a certain velocity draws ions toward it and thereby produces an attractive region that in turn draws a second electron. This electron pairing correlates the motion of all the conducting electrons and maintains perpetual conduction of a current as long as the temperature remains low enough for the pairing not to be broken up. Interestingly enough, the metals most likely to become superconducting at low temperatures are those that are relatively poor conductors at room temperature, and the reason is the very fact that in these metals the electrons interact strongly with the lattice vibrations.

Lattice vibrations can also have a salutary effect on the mobility of carriers in certain semiconductors where most of the time the carriers are held closely to the ions in the lattice. Nickel oxide doped with a trace of lithium is such a material. At room temperature an electric field will cause the carriers to hop occasionally from one side to another, but this happens infrequently. If, however, the temperature is raised, increasing the amplitude of the lattice vibrations and thus bringing ions closer together during a portion of the vibration cycle, the chances of a carrier's hopping from one side to the next are improved; as a result the mobility of the carriers will increase exponentially with temperature.

One of the oddities of semiconductors is that they do not always obey Ohm's law, which states that the electrical resistivity of a material is independent of the strength of the applied field. It turns out that when a strong field is applied to a semiconductor, the conduction electrons are speeded up from their ordinary thermal velocities to such an extent that their energy is vastly greater than the thermal energy characteristic of the temperature of the crystal. This increased electron energy can correspond to effective electron temperatures as high as 5,000 degrees K. when the crystal itself is at room temperature.

As a result the resistivity, which is ordinarily a property of the material at a given temperature, now becomes a property of the material not at its real temperature but at a higher effective electron temperature. Since the effective temperature can be changed by changing the strength of the applied voltage, the resistivity is now dependent on the voltage, in contradiction to the assertion made in Ohm's law.

This effect does not occur in metals because of the abundance of carrier electrons in them. Since there are few carrier electrons in a semiconductor, the resistance of the crystal can be made high enough to support a large voltage. Moreover, the velocities of the carrier electrons in semiconductors are so small with respect to those of the electrons at the Fermi levels in metals that it is harder for them to transfer to the lattice the energy given them by the applied field. This results from the fact that there are fewer quantum-mechanically acceptable velocity states into which the lattice vibrations can scatter the electrons if they are slow.

The relatively simple systems I have described in this article illustrate the new insights that solid-state research has provided into the manifold electrical properties of materials. The properties of some semiconductors are now well understood in quantitative terms, and we appear to be well on the road to a similarly precise understanding of metals. Among the fruits of this deeper understanding is the sophisticated design of solid-state electronic devices, which are becoming ever more complicated mosaics of substances. The field of applications is large and very active; I shall cite only a few of the recent developments.

One is the extensive search for new superconducting materials. Already superconducting magnets are available commercially, and the dream of lossless transmission lines for electricity seems possible for realization in the future. The central problem here is the construction of a material that will be superconducting at a reasonably high temperature. Within the past few months there has been produced a metal alloy—consisting of niobium, aluminum and germanium—that becomes a superconductor at 20 degrees K., the highest transition temperature yet found. This discovery is all the more remarkable because the previous front runner, niobium-tin with a transition temperature of 18 degrees K., maintained its preeminent position for 13 years. The new high in transition temperatures is almost (but not quite) large enough to make possible the use of liquid hydrogen instead of the more expensive helium as the coolant. There are sound theoretical arguments suggesting that 20 degrees is close to the limiting transition temperature for any superconducting material, but it is conceivable that superconductivity at room temperature might be a possibility for organic materials of appropriate design [see "Superconductivity at Room Temperature," by W. A. Little; SCIENTIFIC AMERICAN, February, 1965].

Among the exotic materials exhibiting superconductivity is strontium titanate—a superconducting semiconductor! This somewhat paradoxical nomenclature is actually not a mysterious as it seems; strontium titanate in its ordinary state is a semiconductor, but by decreasing its oxygen content it can be made metallic and superconducting.

For several reasons the electrical properties of organic materials are particularly interesting. It has been shown that the transfer of electrons from molecule to molecule plays important roles in fundamental processes of living organisms. A molecule that has been studied extensively in this context is that of phthalycyanine, a fairly simple structure that resembles the active center of the complex molecule of chlorophyll and certain other biologically active substances. The electrical properties of a number of organic materials have become important in technology. Organic dyes, for example, are used in photography. And organic chemists are investigating the synthesis of polymers with semiconducting and perhaps even metallic properties, which would make possible plastic wiring in houses and plastic transistors in television sets.

Clearly the electrical properties of materials, from pure elements to complex biological substances, depend in a fundamental way on the details of their electronic structures. It is therefore hardly surprising that the efforts of scientists to understand these properties are inextricably tied in to one of the main currents of 20th-century science: the development of quantum mechanics.

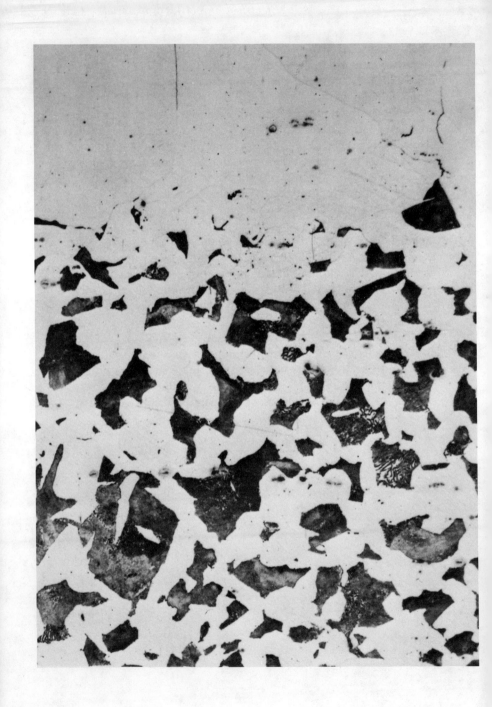

In dealing with solid materials the chemist is concerned not only with such matters as corrosion and chemical syntheses but also with chemical events that occur inside solids; for example, precipitation.

HOWARD REISS

The chemical properties of materials

In the study of the solid state, as in many other fields of modern scientific inquiry, it has become impossible to draw a sharp dividing line between chemistry and physics. Much of the present information about the chemical properties of solid materials has been developed by people trained as physicists, and it is an interesting fact that very few university chemistry departments offer courses in solid-state chemistry as such. Yet every discipline has its own techniques and special folklore, which, by illuminating a subject in a special way, can reveal facets not easily discoverable by other special means. Chemists have brought their own points of view to aid in the study of the physical properties of solids.

Flight of carbon from solid steel is an example of chemical activity within a material. This micrograph shows unalloyed steel, enlarged 300 times. The lower part shows the normal composition of the metal, intimately mixed ferrite and carbon-rich pearlite, formed during heat treatment. In the upper part, close to the sample's surface, the carbon has been preferentially oxidized, transforming the mixture of ferrite and pearlite into ferrite alone. The micrograph was made by Lawrence H. Van Vlack of the University of Michigan and appears in his book Elements of Materials Science. *It is reproduced by the permission of the Addison-Wesley Publishing Company.*

A rough distinction between the physical and chemical approaches to the study of solids might be the following: The physicist is accustomed to working with homogeneous solids in which the specific location of objects or events is of no consequence. This robs him of some opportunity to exercise physical intuition. As a result he transfers his base of operations to an abstract space known as momentum space, in which structure is recaptured. In contrast, the chemist usually deals with localized objects (for instance molecules and elements of structure) in real space. In crystals the chemist is particularly interested in localized defects or imperfections. They are mainly of two general types: point defects and line defects. The point defects include vacancies where atoms are missing, the switching of atoms out of their regular order, the displacement of atoms into interstitial spaces in the crystals, subtitutional impurities (foreign atoms substituted for the regular ones), positive "holes" (where electrons are missing) and atom-like combinations of holes and electrons, called excitons, that can wander through the crystal. The line defects are principally the well-known dislocations so important to the mechanical properties of solids.

Some of the most useful properties of solid materials depend on localized imperfections as well as on the crystal regularity itself. The methods and view-points of chemistry have therefore been important in the development of materials technology by helping to produce defects in a controlled manner. The chemist regards each defect as a species of chemical object localized in space; he largely ignores the regular crystal background as he might the water in typical "wet" chemistry. Viewing crystal defects in this way, he is able to explain many of their effects and interactions on the basis of chemical principles, such as the law of definite proportions and the law of mass action.

As every chemistry student knows, the law of definite proportions asserts that in every sample of a given compound the constituent elements are always present in the same proportions. For example, common salt (NaCl) always contains sodium and chlorine atoms in equal number. Experiments with solid materials have shown, however, that this law can be violated. If we put solid sodium in a closed container with a crystal of sodium chloride and heat the material so that some of the solid sodium is evaporated, we find

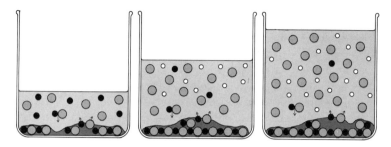

Silver ions are forced out of a solution of silver chloride by application of the "common-ion effect" of the law of mass action. In a beaker (left) *partly filled with a saturated solution of silver chloride and containing a layer of solid silver chloride at the bottom, the system is in equilibrium. Positive silver ions* (black) *and negative chloride ions* (gray) *leave the solid and divide just as frequently as silver and chloride ions in the solution combine and reenter the solid. When sodium chloride is added to the beaker* (center: additional chloride ions are also gray; sodium ions are white), *the formerly pure silver chloride solution has more chloride ions "in common." This increases the frequency of chloride pairing with silver ions, thus reducing the number of silver ions in solution. When more sodium chloride is added* (right), *the number of silver ions in solution is further reduced.*

that the sodium chloride crystal soon acquires more sodium than chlorine. The regularity of the crystal lattice can be maintained only in the presence of the additional sodium ions by the simultaneous addition of vacancies occupied by negatively charged electrons rather than by negative chloride ions.

The defective crystal is said to be "nonstoichiometric." (Adherence to the law of definite proportions is called stoichiometry.) Nonstoichiometry can lead to interesting electrical properties in semiconductors and also to useful optical properties (including coloring) in various materials [see "The Electrical Properties of Materials," by Henry Ehrenreich, page 127, and "The Optical Properties of Materials," by Ali Javan, page 177]. The production of nonstoichiometric defects has therefore become an inviting field for chemists.

Now let us consider the chemical law of mass action as it applies to crystal defects. This law says simply that the rate of any chemical process depends on the products of the concentrations of the interacting species. To illustrate mass action with a simple case in ordinary chemistry, think of a saturated aqueous solution of silver chloride, with some silver chloride deposited in solid form at the bottom of the vessel. Pairs of silver and chloride ions are continually coming out of the solid into solution and, at an equal rate, silver and chloride ions in the solution are pairing up and entering the solid state. This equilibrium will persist as long as the product of the ion concentrations (silver-ion concentration times chloride-ion concentration) remains unchanged. Now add sodium chloride to the solution. The addition of chloride ions increases the product of the silver-ion and chloride-ion concentrations in the solution. With more chloride ions present the chances that silver ions will encounter them are increased. As a result silver-chloride pairs will be deposited on the solid at a higher rate, and the removal of silver ions from the solution will continue until a new equilibrium is established [see illustration on page 147]. Conversely, if chloride ions were removed from the solution, more silver ions would come out of the solid into solution. This "forcing" of a chemical species from one side of a process to the other is known as mass action. Another example is the dissolving of a solid acid by neutralization with a base. When a base such as sodium hydroxide (NaOH) is added to an acid solution in contact with some solid acid, the hydroxyl ions combine with the acid's hydrogen ions (forming water), destroying acid and thus reducing its concentration in solution. This causes more acid to dissolve.

In exactly the same way that the law of mass action controls ordinary chemical reactions, it can be applied to the control of defects in a crystal. We have already noted that any type of defect —a vacancy, an electron "hole" (which behaves like a positive electron), an interstitial atom, an impurity—can be treated in terms of mass action as though it were a chemical entity.

Consider a crystal of sodium chloride. We can think of each sodium ion in its normal position in the lattice as a kind of combination, or "compound," that is capable of generating two offspring: an interstitial ion and a vacancy. By jumping from its normal site into an interstice it produces both defects. A sodium ion adjacent to the vacancy can then jump into it. Therefore at temperatures high enough to agitate ions out of their sites, both interstitial ions and vacancies are able to move through the crystal lattice. Like the elements of a chemical compound, they are subject to the law of mass action, and their concentrations can therefore be controlled by the adjustment of the concentrations of other chemical-like entities.

Suppose we replace a sodium ion in the sodium chloride lattice with a calcium ion. The calcium ion carries two units of positive charge; therefore in order to conserve the crystal's charge neutrality another of the singly charged sodium ions (in addition to the one replaced) must be removed from the lattice. Thus by inroducing calcium ions by adding calcium chloride to the crystal we can create vacancies. The presence of the additional vacancies can result, according to the law of mass action, in a reduction of the number of sodium ions in intestitial positions, so that the product of the two concentrations (of vacancies and interstitials) is restored to its equilibrium value [*see illustrations on next page*].

The law of mass action is applicable even to the energy states of electrons (and holes), which the physicist likes to view from the vantage of momentum space. Some of these states, as they occur in semiconductors, are actually localized imperfections. In terms of energy they are located in the forbidden band, or energy gap, between the valence band and the conduction band. Generated by chemical impurities introduced into the semiconductor crystal, these imperfections are of two types. One comes equipped with an extra electron (attached to an impurity) installed in an energy level just below the conduction band. It is said to be a "donor" because very little energy is needed to donate, or promote, the electron to the conduction band, where it may conduct electricity. Examples of donors in germanium and silicon semiconductors are lithium (an interstitial impurity) and arsenic, antimony and bismuth (substitutional impurities). The other type of imperfection comes equipped with a positive hole in an energy level just above the valence band. It is called an "acceptor" because the hole can easily accept an electron from the valence band, leaving behind a mobile positive hole capable of conducting electricity in that band. Boron, alumnium and gallium are impurities that behave as acceptors in germaniun and silicon semiconductors.

Now, this phenomenon of energy bands and donor and acceptor states allows us to produce a p-n (positive-negative) junction in a semiconductor crystal by the introduction of impurities. This is accomplished by "doping" one region of a silicon crystal with, say,

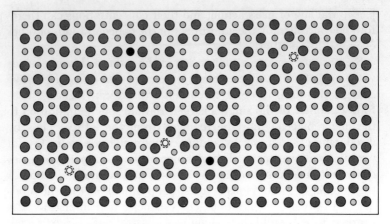

Solid equivalent of the common-ion effect illustrated on page 147 is demonstrated by a salt crystal that is "doped" with calcium chloride. The lattice of sodium ions (dark gray) and chloride ions (light gray) is equivalent to the fluid in the previous illustration. Only the four sodium ions in interstitial positions (broken circles) and the four vacancies at the ions' former sites in the lattice are involved in the demonstration.

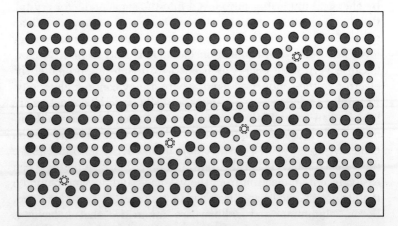

Crystal is doped by adding calcium ions (black) that add vacancies. Just as adding sodium chloride to a silver chloride solution drives silver ions out, so added lattice vacancies can capture interstitial sodium ions. Here one of the four has vanished.

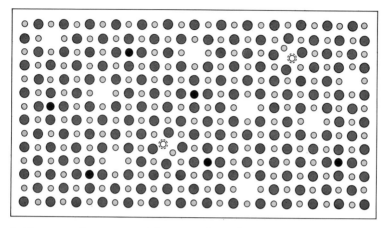

Lattice vacancies are now eight in number, whereas interstitial ions are reduced to two. Their product is identical with the product of interstitials and vacancies before the crystal was doped, thereby meeting the equilibrium requirement of the mass-action law.

arsenic atoms, which behave as donors and yield conduction electrons, while doping an adjoining region in the same crystal with, say, aluminum atoms, which behave as acceptors and yield valence band holes. Because of thermal agitation the mobile electrons and holes wander freely through the crystal. When electrons cross the junction between the donor and acceptor regions, they find holes on the acceptor side and "combine" with them (that is, fill the holes and go out of circulation). Similarly, holes that cross the junction to the donor side combine with electrons. The result of this two-sided process is a net removal of electrons from the donor side and of holes from the acceptor side, so that the donor side becomes positively charged and the acceptor side negatively charged. Mobile electrons are then repelled from the acceptor side, and holes are likewise repelled from the donor side. Accordingly the system comes to equilibrium with a built-in electric field at the p-n junction.

This built-in field makes it possible to fabricate transistors from semiconductors. A single p-n junction can serve as a current rectifier, because an applied voltage can drive current in one direction (holes from the p to the n side and electrons from the n to the p side) but not in the opposite direction, there being no available carriers that can cross the junction for the conveyence of current in that direction. If an n-p junction is paired with a p-n junction, so that an n-p-n region is created in the crystal, the system (then called a junction transistor) can act as a current amplifier.

All of this—not only the production of mobile carriers but also the development of an electric field at a junction—can be discussed in chemical-like terms. The "dissolving" of lithium in solid silicon can be written $Li \rightleftarrows Li^+ + e^-$, with Li^+ standing for the donor ion and e^- for a conduction band electron. Similarly, the dissolving of aluminum is represented by $Al \rightleftarrows Al^- + e^+$, with Al^- signifying the acceptor and e^- a positive hole in the valence band (produced when the acceptor ion traps a valence electron). Now the conduction electron donated by the lithium atom may lose energy and drop from the conduction band into the valence band, there combining with a positive hole: $e^+ + e^- \rightleftarrows e^+e^-$. This reaction exhibits a striking parallel to the process of neutralization of an acid by a base, in which the acid's hydrogen ion combines with the base's hydroxyl ion. We can regard lithium as the counterpart of an acid and aluminum as the counterpart of a base. Then e^+ becomes the analogue of hydrogen ion, e^- the analogue of the hydroxyl ion and e^+e^- the analogue of the water molecule. Following this chemical reasoning, we should expect that, in accordance with the law of mass action, the removal of conduction electrons from the right side of the lithium reaction would shift the equilibrium to the right and cause more lithium to be "dissolved" in the silicon crystal. In short, the presence of the aluminum acceptor increases the "solubility" of the lithium donor. This is the same effect by which a base dissolves an acid through neutralization.

Another way to view this effect is in terms of the built-in field at a p-n junction. Acceptors on the negative side of the junction exert an electrical force that attracts the donors from the positive side. Accordingly the solubility of the donor is increased by the presence of the acceptor.

Experiments in the doping of semiconductors confirm with remarkable precision predictions based on the law of mass action. In fact, agreement between theory and experiment actually turns out to be better in these solid materials than it is in water solutions!

Mass-action effects have been found in insulators and metals as well as in semiconductors, and they apply to all kinds of imperfections, including the defects in nonstoichiometric crystals and even to dislocations. A dislocation may contain, along its length, atomic configurations that can be ionized; thus a linear dislocation is analogus to the chainlike protein molecule in solution, whose chemical groups can also be ionized. It has been found that the state of charge on a dislocation affects its ability to move through the crystal; hence it becomes apparent that any impurity or other defect that influences the state of charge can have a pronounced effect on the mechanical properties of the material.

Another interesting chemical phenomenon that is demonstrable in solids is the "diffusion-controlled" reaction. This concept arises when the rate of a chemical reaction depends primarily on the time required for the reacting species to diffuse to one another. Chemists have studied diffusion-controlled reactions in liquid solutions for many years. One example is the quenching of fluorescence in a liquid solution. Certain substances in liquid solution, after being raised to an excited energy state, give up their energy spontaneously in the form of light. This fluorescence is forestalled, however, if the excited molecules happen to encounter other molecules (quenchers) that absorb the excess energy in the form of heat before they can lose it in the form of light. Chemists have been able to calculate the rate at which such quenchers diffuse to the excited molecules. The mathematics of diffusion-controlled reactions has been worked out in considerable detail, and it turns out that the rate of these reactions is influenced considerably by what might be called the "experience" of the excited molecules. This means that a molecule that persists without being quenched gains information about the location of quenchers with respect to itself.

Consider a crystal of copper that has been damaged slightly by bombardment with a beam of electrons of moderate energy. The bombardment displaces a number of the copper atoms from normal sites into interstices in the lattice. Gradually the damage will heal as the mobile interstitial atoms and vacancies, diffusing through the lattice, meet one another and recombine. Hence we have a diffusion-controlled reaction in a solid. The molecules' encounters, however, are not an entirely random affair. Vacancies that survive for a considerable time do so because there were few interstitial atoms in their neighborhood. These vacancies therefore gained some "information" about the distribution of interstitial atoms in their vicinity, so that their recombination is not with a random environment. This "knowledge" on the part of the reacting species plays a part in the rate of recombination.

The physicists who studied the healing of radiation damage in crystals dealt with the matter at first as if the rate of healing depended simply on the law of mass action in a straightforward way, ignoring the nonrandom aspects mentioned above. Collaboration with chemists who were familiar with the effect led to the proper corrections. This is an instance in which the combination of chemical and physical approaches provided a more satisfactory solution to a problem than either approach did separately.

It should not be supposed that chemical-like reactions in solids are generally of such an esoteric nature. More familiar reactions occur. Precipitation, for example, is a common phenomenon within solid materials. Precipitates usually take the form of small particles—aggregates of atoms that come out of solution in the solid and are deposited at various places in the crystal. They can exert a strong influence on the properties of the solid.

If to a solution of lithium in solid silicon one adds another electron-donating substance, the lithium's solubility will be reduced. Lithium atoms therefore precipitate out of solution by diffusing through the crystal until they collect in sizable aggregates (perhaps containing some silicon atoms as well). The formation of the aggregates is a diffusion-controlled reaction. Another interesting precipitate occurring in silicon involves oxygen. Precipitated oxygen atoms form aggregates by joining with silicon atoms in chainlike "polymers." Some of these aggregates are electrically active and donate conduction electrons to the crystal. In the early days of the silicon transistor an unwanted excess of conduction electrons sometimes developed mysteriously; the phenomenon was eventually traced to the presence of precipitated oxygen.

In addition to influencing the electrical properties of semiconductors, precipitates can produce other effects in solids. For example, they can alter the brittleness or strength of a crystalline material by interfering with the movement of dislocations.

Solids can be the seat of another chemical process that is better known in connection with liquids. This is the familiar wet-cell electric battery phenomenon. Such a battery might consist of a zinc plate immersed in a water solution of zinc sulfate and a copper plate in a solution of copper sulfate, the two solutions being placed in contact. Atoms from the zinc plate dissolve in the zinc sulfate as positive ions. The electrons left behind on the solid zinc travel by way of an external circuit to the copper plate; there they combine with positive copper ions from the copper sulfate solution, and the copper atoms thus formed are deposited as copper metal. The circuit is completed by the movement of negative sulfate ions from the copper side to the zinc side of the solution, thereby balancing the added zinc ions. The overall chemical reaction is simply the reaction of zinc with copper sulfate to produce copper and zinc sulfate. The sulfate solutions serve as electrolytes for the transport of the current, and energy to drive the current is provided by the chemical reaction.

Now, this type of process can take place in a system with a solid compound, instead of a liquid solution, serving as the electrolyte. Consider an arrangement in which the electrodes are solid silver on one side and carbon in contact with bromine gas on the other, and the electrolyte is solid silver bromide. Silver ions can diffuse rather rapidly through solid silver bromide. As atoms from the silver electrode dissolve in the silver bromide (becoming positive ions) the electrons remaining behind in the silver electrode travel by way of an external circuit to the carbon. Meanwhile silver ions, having crossed to the carbon through the bromide, combine with the electrons and the bromine gas to form silver bromide. Hence the circuit is completed: current is carried by electrons in the external part of the circuit and by positive silver ions in the electrolyte. Note that the electrolyte must be a nonconductor of electrons; if electrons could travel through it, they would cross from the silver electrode to the carbon by that route and thereby short out, or bypass, the external circuit.

Again, as in the wet sulfate battery, the solid-state version is energized by a simple chemical reaction: the combination of silver with bromine to form silver bromide (a solid that does not conduct electrons). Solid-state batteries are now receiving a great deal of attention, particularly for use in space vehicles and possible use in automobiles. Among the encouraging recent developments is the discovery of solid electrolytes in which ions can diffuse about as rapidly as in some liquid electrolytes. These discoveries were made independently and almost simultaneously at the University of Essex in England and at North American Aviation, Inc.

Still another important field of research on solid-state materials is their surface chemistry. Most solid surfaces, even when they seem very smooth, are riddled with submicroscopic defects—atomic vacancies, kinks, ends of dislocations, dangling chemical bonds and so forth—that make it easy for foreign atoms to become attached to the surface. Layer on layer of contaminating material may build up in this manner. A familiar example, of course, is the corrosion of metal surfaces. When an oxide film covers a metal, it is likely to contain oxygen-atom vacancies or metal atoms displaced into interstitial spaces or both. Oxygen atoms can be added to the oxide lattice at the oxygen-air interface, forming metal vacancies in much the same way that sodium added to sodium chloride forms chloride vacancies. Metal atoms can jump into these vacancies and by this process are able to diffuse through the film. Thus metal from the interface between metal and metal oxide is transported to the air interface, where it combines with oxygen to form more oxide.

Electrochemical mechanisms exist that can inhibit the oxidation of a metal surface. These have been studied by immersing a metal electrode in an oxidizing electrolyte solution and applying a voltage to the electrode; when the flow of current falls off, it signals a decline in the rate of oxidation. This phenomenon has often been observed; the electrochemical reaction begins at a rapid rate and then slows substantially. Two mechanisms that can bring about such a state of "passivity" have been identified. When a metal electrode is immersed in moderately concentrated nitric

Epitaxial growth of silicon on a substrate is seen in two intermediate stages in these micrographs. The sample at the top, derived from an extremely dilute silane source, shows early growths that exhibit crystallographic symmetry. The early growths soon coalesce into larger, unsymmetrical islands. The sample at the bottom, derived from a silicon tetrachloride source, has initial nuclei shaped like small pyramids. The circles outline the traces left by some of the pyramids after they were absorbed by the adjacent islands. Both replica electron micrographs were made by R. L. Nolder of the Autonetics Division of North American Aviation, Inc.

Chemical properties of materials 157

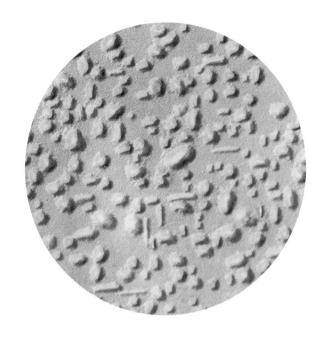

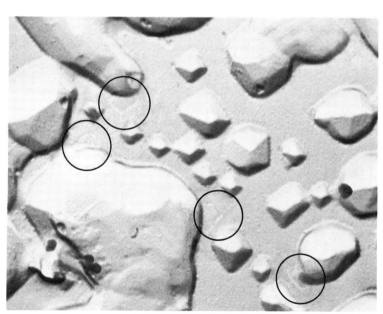

acid (HNO_3), the acide molecule, reacting with an electron, gives up a negative oxygen ion (and becomes nitrous acid: HNO_2). Consequently a layer of negative oxygen ions is laid down on the metal surface; these pull positive metal ions from the surface and form a layer of oxide. As the film grows it becomes difficult for metal atoms to pass through the film to the oxygen, and the oxidation therefore falls to a negligible rate. Another passivating mechanism involves carbon monoxide. If a metal electrode is immersed in a solution of hydrochloric acid containing carbon monoxide, a layer of carbon monoxide is often adsorbed on the metal surface. In this case, however, the carbon monoxide layer apparently has no tendency to grow but acts as a film that electrically insulates the metal from the electrolyte solution. The electrochemical reaction therefore comes to a halt and the metal is "passivated."

Surface defects are believed to account for the reaction-promoting properties of some solid catalysts. One hypothesis concerning the mechanism of catalytic activity, which is supported by some experimental evidence, assumes that the molecule about to undergo chemical reaction is adsorbed on the catalyst surface by the attachment of its atoms to dangling chemical bonds. As a result of this attachment some bonds within the molecule are weakened and break as the molecule becomes detached from the catalyst. The molecule thereupon splits in two and completes the reaction. Such a mechanism depends, of course, on a fit between the configuration of the molecule and the atomic arrangement in the catalyst's surface. It also depends on the strength with which the molecule is adsorbed on the surface: if the adsorption is too strong, the molecule will simply remain attached; if it is very weak, there may be no catalytic action. A reaction that is thought to proceed by this mechanism is the removal of hydrogen atoms from the hydrocarbon cyclohexane over a zinc oxide catalyst.

Surface geometry plays a key role in another surface process. This is the process in which one crystal is grown on another—the phenomenon known as hetero-epitaxy [*see illustration on page 157*]. For example, the geometric arrangement of atoms at the surface of beryllium oxide is such that single crystalline layers of silicon atoms can be laid down on it. Such an arrangement, with a semiconductor laid on an insulating base, is used in the fabrication of compact integrated circuits.

A particularly interesting phenomenon is the ability of doped semiconductors—for example a p-type material—to act as catalysts. The more valence band holes there are in the solid, the more effective is its catalytic action. This implies that some step in the catalytic process involves the donation of electrons from the reacting system to the catalyst. The transfer of electrons may occur during the process of adsorption or during the chemical process itself. In any event, the phenomenon is interesting because it illustrates how a bulk property of a catalyst (the degree of its p-typeness) can influence a surface reaction. Doping the catalyst to make it more or less p-type (that is, controlling the concentration of electrons) can change its properties.

At the surface of a semiconducting catalyst there is usually a built-in electric field very similar to the field at a p-n junction. Work must be done to move holes or electrons through this field, and this shows up in the energy required to advance the catalyzed reaction. Indeed, energy considerations dominate all chemical reactions and the physical properties—electrical, magnetic and optical—of all materials.

Why atoms are magnetic is well understood, but why some materials are magnetic is less so. Nonetheless, advances in magnetic materials have made possible devices from refrigerator latches to computer memories.

FREDERIC KEFFER

The magnetic properties of materials

Magnetic materials provide a good illustration of how man's practical reach can exceed his theoretical grasp. The peculiar properties of magnets have been known since lodestones were recognized 3,000 years ago. Every day in the U.S. alone magnetic materials are used to generate three billion kilowatt-hours of electric power. Magnetism also works to distribute electric power efficiently, to energize electric motors, to reproduce sound and visual images, to store information, to latch doors and turn speedometers. Until 40 years ago, however, no one understood why certain materials were magnetic. Even today there are more basic questions about magnetism than there are answers.

Almost everything that is known about the magnetic properties of materials has been obtained from experimental discoveries—

Magnetic domains appear as a pattern of zigzags on the surface of an alloy of manganese and bismuth in the micrograph shown here. The domains are visible because the plane of polarized light reflected from a magnetized surface is rotated in proportion to the degree of magnetization. The micrograph was made by the metallographic unit at the Research and Development Center of the General Electric Company. Other domain micrographs, in which a magnetic powder outlines the domain walls, appear on pages 170 and 172.

some intentional but many the result of lucky accidents—and from a few bold inductive insights. Little progress has been made deductively. Although no one seriously doubts that all magnetic properties flow naturally from quantum mechanics and electromagnetic theory as it is applied to many-atom systems, the systems themselves have generally proved to be too complex for analysis by reasoning from the general to the particular.

Three basic magnetic properties of materials have called for explanation since they were first noticed. The first is the fact that some materials are magnetic in the absence of any applied magnetic field. Associated with this property is the fact that the same materials generally become more magnetic when a comparatively weak magnetic field is applied to them. This dual property is what we call ferromagnetism. The second property, which is characteristic of the same materials, is the disappearance of the ferromagnetism when the material is heated above a specific temperature. Above that temperature ferromagnetism is replaced by a comparatively weak magnetism that varies in proportion to the intensity of any applied magnetic field and is along the direction of the field. This property is termed paramagnetism. The third property that requires explanation is exhibited by *all* materials and is a magnetic response in a direction opposite to that of any externally applied field. This property, called diamagnetism, is much too weak to be of any practical value; it is scarcely noticeable in materials that are ferromagnetic or paramagnetic.

The lodestone, which is the mineral magnetite (Fe_3O_4), is the principal natural magnet. It is presumably magnetized when it cools from the molten state in the presence of the earth's magnetic field, even though that field is extremely weak (about .5 oersted). Metallic iron, however, is the magnetic material par excellence. Freshly smelted iron is rarely magnetic, but if the iron is drawn into the shape of a bar (as William Gilbert demonstrated at the end of the 16th century), the earth's magnetic field will magnetize it. So will any other magnetic field, such as a lodestone's, and the bar will remain magnetized after the field has been removed. The critical temperature at which the spontaneous magnetism of magnetite changes to paramagnetism is 585 degrees centigrade; the critical temperature for iron is 770 degrees C. The critical temperature of a ferromagnet was named the Curie temperature after Pierre Curie, who was a pioneer student of paramagnetism.

In 1907 another French physicist, Pierre Weiss, provided the first inductive insight that came anywhere near accounting for the behavior of magnets. Even those scientific workers who frequently accept the role of speculative leaps might boggle at Weiss's flight of fancy. There appeared to be not the slightest theoretical justification for two new concepts he introduced: the "domain" and the "molecular field."

According to Weiss, a ferromagnet that is below its Curie temperature is made up of small, permanently magnetized regions — domains. The total magnetic strength of the material is simply the sum of the magnetic strengths of all its domains. If the magnetic axes of the various domains happen to point in many different directions, the material's total magnetic strength will be very small, even zero. The application of an external magnetic field, Weiss held, aligns the magnetism of the domains much as the wind will align a group of weather vanes, thus greatly increasing the material's total magnetic strength. The applied field need not be particularly strong; a light breeze turns weather vanes as effectively as a gale does. Weiss visualized his domains as being small but macroscopic. We now have direct evidence that the domains exist; they are usually between .1 and .01 centimeter across.

Compared with the dimensions of atoms and molecules, Weiss's domains were enormous. Accordingly they were not closely related to any particular theory of atoms and molecules. His second concept — the molecular field — was intended to explain at the atomic level why domains are present in ferromagnets below the Curie temperature but absent above that temperature. Weiss rested his molecular-field hypothesis on the assumption (prevalent at the time but not confirmed by experiment until much later) that the atoms of certain materials were themselves infinitesimal magnets. Below the Curie temperature, Weiss proposed, an intrinsic magnetic field kept these atomic magnets in alignment. Above the Curie temperature thermal agitation was strong enough to overcome the aligning power of the intrinsic field. The material then lost its ferromagnetic properties and exhibited paramagnetism only.

The intrinsic field that Weiss postulated was not only strong but also mysterious. He assumed that it was related in some way to the material's state of organization at the atomic or molecular level; hence his name for it. The thermal energy required to overcome the field could be estimated from the temperature at which it disappeared, and so its strength could be estimated. For iron this proved to be on the order of 10 million oersteds—greater than the strongest steady magnetic fields produced today with many megawatts of power.

Weiss, of course, had not the slightest notion of the origin of the hypothetical field. The mystery remained unsolved until 1928, when Werner Heisenberg of Germany and Y. I. Frenkel of the U.S.S.R. independently suggested that the field probably arose from large electrical forces at the atomic level that were masquerading as magnetic forces because of a subtle quantum-mechanical effect. Hence both of Weiss's hypotheses, proposed well in advance of any supporting theory or evidence, have proved to be generally correct.

Why is it that some, rather than all, atoms are infinitesimal magnets? Today we know that atomic magnetism is almost entirely the result of an imbalance of an atom's electrons. Some atomic nuclei are also little magnets, but because the strength of their magnetism is inversely proportional to their mass it is weaker than the magnetic strength of electrons by a factor of about one to 2,000. Electron magnetism is of two kinds: the magnetism associated with the angular momentum of the electron's spin and the magnetism associated with the angular momentum of the electron's orbital motion around the nucleus. In either case the direction of the resulting magnetism is along the axis of rotation.

In most kinds of atoms—or at least in most assemblages of atoms in molecules and solids—there is a tendency for both the spin and orbital angular momentums of adjacent electrons to cancel each other by the formation of antiparallel pairs. In such a pair the spin of one electron, for example, will be clockwise and the spin of the other counterclockwise. Their total momentum is therefore zero, and so is their total magnetism. It is this tendency—or rather its slight variability—that accounts for the extremely weak magnetism of "nonmagnetic" materials. When the application of a magnetic field slightly unbalances the orbital pairing of electrons, the

Magnetic properties of materials 165

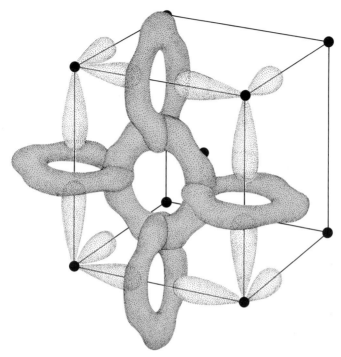

Lattice cell, one portion of an iron crystal, consists of nine atoms (black) in a cubic array, one at each corner and one in the center. Stippling in light gray indicates areas of strong positive magnetization associated with the unpaired electrons in the four atoms that form the face of the cube. Stippling in dark gray indicates areas of weak negative magnetization along the face of the cube. Positive and negative areas elsewhere are omitted. C. H. Shull of Massachusetts Institute of Technology and H. A. Mook of Harvard University determined the distribution of the areas by scattering magnetically polarized neutrons from an iron target.

result is diamagnetism; when it slightly unbalances the spin pairing (which it can do only in metals), the result is a very weak form of paramagnetism.

The Pauli exclusion principle requires that no two electrons occupy the same state, by which is meant both the same region of space and the same spin and orbital momentums. It is possible, however, for the two electrons to occupy the same region if they are an antiparallel pair. In quantum mechanics the stricture against occupying the "same" region simultaneously is overcome because the two electrons behave like superposed, smeared-out charge distributions. It is this behavior, together with the attractiveness of the nucleus for electrons, that gives rise to the antiparallel pairing tendency of all atoms. It rather resembles the way in which space is saved in a sardine can by packing the fish head to tail (except that even more space could be saved if antiparallel sardines, like electrons, completely interpenetrated one another).

The antiparallel tendency must, of course, overcome the mutual repulsion of the two interpenetrating electrons. It usually does. The outstanding exceptions are found in the "transition" metals and the rare earths. These elements are characterized by atoms in which some electrons have not formed antiparallel pairs. In reality any atom that has an odd number of electrons is magnetic, and many kinds of atoms have outer electrons that are unpaired. These outer electrons, however, usually form antiparallel pairs with the outer electrons of neighboring atoms on coming together in a solid. What gives the magnetic atoms in the two groups of elements their unique property—and thereby gives rise to strongly magnetic materials—is that the unpaired electrons are located not in the outermost electron cloud but in the inner ones, where they are not able to form pairs with electrons in other atoms.

The atoms of ferromagnetic materials are tiny magnets because of their unpaired inner electrons. These electrons, however, are more or less screened from the adjacent atomic magnets by outer electrons. How, then, can we account for ferromagnetism? In a ferromagnetic material neighboring atoms tend to be magnetically aligned. The picture is complicated in its details, but it is evident that Weiss's concept of a long-range molecular field, coupling each atomic magnet with equal strength to all other atomic magnets, was partly mistaken. Instead the coupling effect is a short-range one, being sharply limited by electron screening.

Short-range coupling nonetheless exerts a powerful influence. Consider the example of a ferromagnetic material that has been heated above its paramagnetic threshold and then allowed to cool. As it approaches its Curie temperature, the energy of thermal agitation gradually becomes less than the coupling energy. At this point each atomic magnet begins to line up with its nearest neighbors, and those with their nearest neighbors and so on, as if all the atoms were being guided by the kind of long-range field visualized by Weiss. One is reminded of the situation when, as the quiet of evening descends, suddenly all the dogs in a town get to barking together, although each dog responds only to the neighboring dogs.

The coupling effect can also produce antiparallel alignments. Manganese fluoride, for example, is paramagnetic at temperatures above minus 206 degrees C. Below this transition temperature (which is analogous to the Curie temperature of a ferromagnet) each manganese atom that, so to speak, points upward is paired with one that points downward, and no overall magnetism is observed. This property is called antiferromagnetism. The transition temperature between the antiferromagnetic and the paramagnetic state has been named the Néel temperature after the French physicist Louis Néel, who with his colleagues first clarified the phenomenon.

There are other coupling arrangements, the most notable of which produces ferrimagnetism (not to be confused with ferromagnetism). When magnetite is cooled below its Curie temperature, for example, the short-range coupling is such that two out of the three iron atoms in the material are aligned by pointing upward and the third points downward, so that there is a net magnetism due to one atom in three. Thus the oldest of magnets is not a ferromagnet at all. The macroscopic properties of ferrimagnets, however, are generally similar to those of ferromagnets. Among the man-made ferrimagnets are the technologically important ferrites and iron garnets. The ferrites were developed in the 1940's by J. L. Snoek and his associates at the Philips Research Laboratories in the Netherlands. The iron garnets were first made in the late 1950's by Néel's group in France and by workers at the Bell Telephone Laboratories.

Coupling forces that favor antiparallel alignments sometimes extend an appreciable distance, to the point of affecting neighboring atoms twice removed. Since it is obviously impossible for each atomic magnet to line up antiparallel to very many distant neighbors, some second-nearest and third-nearest atomic magnets tend to end up parallel with respect to one another in spite of the coupling energy. One compromise, resulting in a stable situation, is made when adjacent atomic magnets form a helix. In a crystal of manganese dioxide below a temperature of minus 189 degrees C.,

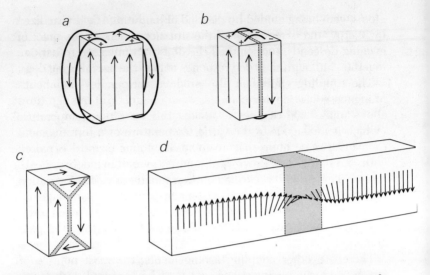

Domain formation is caused by the spontaneous subdivision of a uniformly magnetized material (a). Division into two domains (b) lowers the external field energy of the material, as is evident from the shortened external magnetic lines of force in this example. The next division, from two to four domains (c), eliminates all external lines of force. The zone between domains, known as a Bloch wall, is shown in detail at bottom (d). In this zone some "north" poles in the material, which are turning through 180 degrees, come to the surface of the material and cause the lines seen in the photograph on page 170.

for example, there is a turn angle of nearly 129 degrees in each plane of atoms, so that the helix makes five revolutions in the process of crossing 14 planes. This compromise property, known as helimagnetism, was discovered simultaneously in 1959 by A. Yoshimori of Japan, J. Villain of France and T. A. Kaplan of the Lincoln Laboratory of the Massachusetts Institute of Technology. The existence of helimagnetism immediately suggests an interesting generalization. Antiferromagnetism, for example, could be defined as helimagnetism with a turn angle of 180 degrees and ferromagnetism as helimagnetism with a turn angle of zero degrees!

It is relatively easy to see how the domain hypothesis explains ferromagnetism. Not until some three decades ago, however, was it realized how domains too could be explained by means of a fundamental principle of physics. This is the principle that at low temperatures all systems tend toward a state possessing minimum energy. The external lines of force of a uniformly magnetized material are shrunk and the energy of the external magnetic field is reduced if the sample is subdivided by increasing the number of its domains [see illustration on facing page]. The Russian physicists L. D. Landau and E. M. Lifshitz were the first to note this effect in 1935. They pointed out that, whereas the multiplication of domains within a ferromagnetic material would reduce the energy of the external field, it would simultaneously raise the short-range coupling energy, because the number of atomic magnets with antiparallel alignment along domain boundaries would be increased. Eventually the process of subdivision reaches a state of equilibrium; the addition of further domains would increase the coupling energy more than it would reduce the field energy.

The antiparallel alignment of the atomic magnets on opposite sides of a domain boundary is not abrupt but a gradual transition that occupies a zone named the "Bloch wall" after Felix Bloch, who investigated domain boundaries in the 1930's. The "turnover" within the Bloch wall is much more gradual than the twisting in helimagnetism; the zone is some 300 atomic planes thick compared with a few planes at most. Both helices, however, are the result of a similar compromise in the sharing of coupling energy.

In 1931 the late Francis Bitter, working at the Westinghouse Research Laboratories, proved the existence of domains by making them visible. As later refined, the Bitter technique consists of polishing the surface of a magnetic material, spreading a colloidal suspension of magnetic iron oxide particles over the surface and placing a microscope cover glass on top. The particles collect along the lines where the atomic magnets in each Bloch wall point toward the surface, making the domain boundaries plainly visible through a microscope [see upper illustration on page 170]. The shifts in domains under the influence of external magnetic fields that could be observed with Bitter's technique made it possible to investigate such properties as reversible and irreversible magnetization.

From the technological viewpoint an ideal magnetic material should possess one or the other of these properties. It should either be "soft," meaning easy to magnetize and demagnetize, or "hard,"

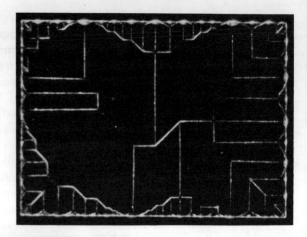

Domain walls in a single crystal of nickel are made visible under the microscope by a coating of magnetized iron oxide. The powdered oxide gathers along the lines at which poles in the Bloch wall point to the face of the crystal (see illustration on page 168).

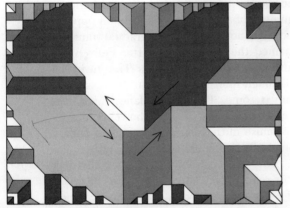

Analysis of domains in the nickel crystal in the upper illustration is shown in this diagram. Most domains prove to have one or another of four directions of magnetization, comprising two sets with opposite polarity (arrows). Both the micrograph and the diagram were prepared by R. W. DeBlois of the General Electric Research and Development Center.

meaning the opposite. Hard materials come into play whenever permanent magnets are required; soft ones are needed for electric generators, motors and transformers. There was a practical means of telling soft and hard materials apart long before there were theories explaining the two properties. A material was placed in a magnetic field, the intensity of the field was varied and the resulting magnetization of the material was plotted as a curve.

Using this technique in 1885, the British physicist J. A. Ewing noted that after the external field was reduced to zero the specimen continued to show net magnetization. When Ewing reversed the field, he found that net magnetization in the reverse direction persisted after the field was reduced to zero. He named this tendency for induced magnetization to remain after the applied field was removed "hysteresis," after the Greek word meaning "to lag."

Any generator, motor or transformer would operate with maximum efficiency if no magnetization remained after the applied external field dropped to zero. Because of hysteresis a vast treasure is wasted each year. The underlying cause of the phenomenon has been revealed by the study of domains.

When a weak external field is applied to a magnetic material, the material's domain walls are seen to move. The motion is in a direction that increases the size of the domains whose magnetization is parallel to the applied field and reduces the size of domains whose magnetization is antiparallel. These increases and reductions are reversed when the applied field is reversed. When a stronger external field is applied, however, the domain walls are often pushed past obstacles, thereby rendering some of the specimen's magnetism irreversible. Obstacles exist at the boundaries between the material's crystal grains, at places where nonmagnetic inclusions are present and at imperfections [*see illustration on page 172*].

Irreversible magnetism is trapped behind an energy barrier, so to speak, when the external field is reduced to zero. It cannot be changed unless an external field of opposite polarity and exceeding some threshold is applied. If the energy state in which the material has been left is not the lowest state possible, the system is said to be metastable. Most ferromagnets contain imperfections that act as obstacles. As a result they exist in a large number of metastable states, and the immediate behavior of any ferromagnet depends markedly on the particular metastable state into which it was last maneuvered. This is one reason why the study of magnetic materials, if not an art, is certainly a science that depends heavily on good luck.

Irreversible magnetization means that the plot of a material's demagnetization path does not overlap its magnetization path. The figure described by the two paths is known as the hysteresis loop; it is narrow for soft materials and wide for hard ones [*see illustration on page 173*]. The inexpensive fabrication of soft mag-

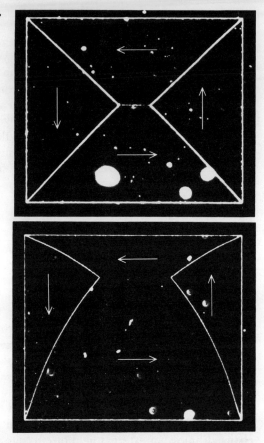

Wall motion in a magnetized material is produced by application of an external magnetic field. The domain walls in a single crystal of nickel-cobalt alloy (top) *have been shifted* (bottom) *by applying a 3.6-oersted field favoring the growth of the lower domain.*

netic materials is difficult because the ferrous materials that are most easily made are full of such obstacles as crystal boundaries and randomly oriented crystal grains, all of which give rise to hysteresis. The ideal soft magnet would be a ferrous material that was cheap to make and in which the crystal grains were all oriented in the same or nearly the same direction.

In 1934, after thousands of experiments, N. P. Goss of the Cold Metal Process Company perfected an orientation process using steel with a 3 percent content of silicon. He subjected the steel to

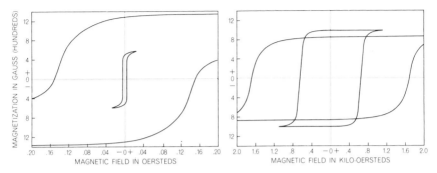

Extreme difference between "soft" (left) *and "hard"* (right) *magnetic materials requires an increase by 10,000 in the horizontal scale of graph at right if hysteresis loops typical of both are to be compared. The large loop at left is for Goss 3 percent silicon steel; the very small one is for supermalloy. The smaller loop at right is for Alnico No. 5. The much larger loop is for Alnico No. 9.*

a moderate cold-rolling followed by an annealing, another moderate cold-rolling and a high-temperature annealing. The theoretical basis of the very exacting Goss recipe is not well understood. It appears that impurities, usually manganese sulfide, keep all but similarly oriented crystal grains from growing and that the pinch of sulfur is critical—there can be neither too much nor too little.

The percentage of silicon in the recipe represents a compromise. The addition of silicon increases steel's electrical resistivity (usually a desired characteristic) but makes the metal brittle. Where mechanical strength is required, as in the rotating parts of generators and motors, steels that contain 1 to 3 percent silicon are used. A higher silicon content can be tolerated in transformers, but steels with more than 4.5 percent silicon are at present too brittle for normal use.

Many metallurgists are nonetheless working toward a reasonably ductile 6.5 percent silicon steel. At this percentage, for some entirely mysterious reason, the metal loses the property of magnetic materials known as magnetostriction. An electrostatic interaction that physically stretches each iron crystal slightly in the direction of magnetization, magnetostriction adds to hysteresis losses. It also appears to be the reason transformers make a humming sound.

An improvement over Goss steel, made according to an even more exacting formula, is "cube-textured" steel. In this material the cubic face of each crystal grain lies parallel to the surface. This minimizes the energy spent in overcoming another source of hysteresis loss, unfavorable anisotropy: a nonsymmetrical distribution of the crystal's energy field.

In some applications the cost of a material is secondary to the reduction of hysteresis loss. An example is the inductive loading of submarine cables. Permalloy, a mixture of 78.5 percent nickel and 21.5 percent iron in which both magnetostriction and anisotropy are minimal, was perfected for cable telegraphy. The anisotropy of iron is opposite in direction to that of nickel and the two nearly cancel each other. Subsequent research showed that adding a third element to the alloy—usually molybdenum—can drive both magnetostriction and anisotropy to zero. The improved combination has been christened supermalloy. Consisting of 79 percent nickel, 15.7 percent iron, 5 percent molybdenum and .3 percent manganese, it shows an extraordinary ability to become magnetized in almost negligible fields.

Ferromagnetic materials exposed to changing external fields are subject to energy losses besides those caused by hysteresis. One such important loss in conducting materials arises from eddy currents, which are induced by field changes and which increase as the frequency of the charges increases. Any enhancement of electrical resistivity—for example by the introduction of impurities—will lower eddy-current losses. The reduction of the area through which the currents travel also cuts losses; this is done by building the magnet out of alternate layers of metal and insulation.

Some ferrites have an electrical resistivity 100 billion times greater than the resistivity of metals, which makes them immune to eddy currents. As a result these ferrimagnetic materials are often used in high-frequency devices. Their major application at present is in the "flyback" transformers of television picture-tube scanning systems, but ferrites also have important uses as elements in microwave circuits. Because not all the atomic magnets in ferrimagnetic materials point in the same direction, the maximum magnetization of ferrites is limited. They are therefore ruled out for most generator, motor and transformer applications.

Logically enough, the first efforts to produce hard magnetic materials for industrial use emphasized the addition of various internal obstacles. One early process involved the rapid cooling of steel containing about 1.2 percent carbon. Iron carbide normally starts to precipitate at 870 degrees C.; rapid cooling of the steel kept the finely divided material suspended throughout the crystal lattice, which also underwent many internal strains because of the sudden reduction in temperature. The process brought about a "martensitic" transformation in the steel. Unfortunately martensitic steel is magnetically unstable. Its magnetization is easily altered by shock, vibration or even variations in temperature. Instability remains a disadvantage in the vastly improved martensitic alloys perfected in the 1920's by the Japanese metallurgist K. Honda.

Recent years have seen the development of an entirely new class of permanent magnets. The theory underlying them was first stated by Néel, who found that a small enough particle of magnetic material cannot contain more than a single domain and thus cannot support any Bloch walls. In iron, he found, the critical size was a diameter of roughly .000002 centimeter. It has since been demonstrated that elongated single-domain particles, when they are magnetized in the direction of their elongation, are magnetically extremely stable.

The first, and still the most widely used, of these materials are the precipitation alloys, of which Alnico is probably the best known. T. Mishima of Japan invented the prototype and called it MK steel; its formula was approximately Fe_2NiAl. G. D. L. Horsburgh and F. W. Tetley of Britain further improved the alloy by adding cobalt and copper. Alnico magnets are fabricated in columnar form by pouring the molten alloy into a cylindrical mold with a cold bottom and hot walls. The upward freezing of the metal produces elongated particles. At the same time the alloy decomposes into a magnetic component rich in iron and cobalt and a nonmagnetic component rich in nickel and aluminum.

Some ferrites with very large anisotropy have also proved to be good permanent magnets. One of them, marketed by the Westinghouse Electric Corporation under the name Westro, is made by aligning the individual crystals in powdered strontium ferrite, usually by means of a magnetic field, and then pressing and sintering the powder. This ceramic material has found many applications and has even been mixed with rubber to make flexible magnets, such as gaskets for refrigerator doors.

Another elongated, single-domain material is marketed by the General Electric Company under the name Lodex. Iron alloyed with 30 or 35 percent cobalt is electrodeposited in mercury. The particles are removed from the mercury, coated with antimony and placed in a lead matrix that is then ground into a powder. Each grain of the powder is a tiny magnet. The softness of the lead matrix allows the material to be cold-pressed and stamped into unconventional shapes that find application in such dissimilar devices as speedometers and hearing aids.

One application of magnetic materials calls for properties that lie halfway between hard and soft. The magnetic-memory element in a digital computer must be hard enough to retain its forward or reverse magnetization—the states corresponding to 0 and 1 in the binary system—indefinitely on being stored. It must also be soft enough to switch states cleanly and rapidly when a small external field is applied in the course of information read-in or read-out. Tiny ferrite cores are being replaced today by thin magnetic films in the interest of faster switching.

William Gilbert wrote in 1600 that it was "by good luck" that "smelters of iron or diggers of metal" had discovered magnetite perhaps as early as 800 B.C. Luck has played no less central a role in the development of the theories by which we explain magnetic properties. In recent years systematic application of these theories has begun to yield planned materials. The recipe still calls for a large measure of luck, however, and will no doubt continue to do so.

The quantum-mechanical interpretation of the spectroscopic characteristics of the elements has made possible a number of technological advances such as the development of lasers.

ALI JAVAN

The optical properties of materials

Before the advent of quantum physics one could give only very crude answers to such elementary questions as: Why do materials have characteristic colors? Why do all materials glow when they are heated? What makes one material transparent and another opaque?

We now know that all these optical properties are intimately related to the way electrons are deployed in a material. This chapter explains how the modern theory of quantum mechanics accounts for the optical properties of materials in terms of their electronic structure, and shows how this knowledge is being used in the development of new materials for optical applications. For historical as well as expository reasons I shall begin by describing the optical behavior of isolated atoms, such as one finds in a gas, and then proceed to show how this comparatively simple picture is related to the complex many-body problem presented by a typical solid material.

The insights of quantum mechanics have come in large part from spectroscopy, which began more than 300 years ago with the experiments of Isaac Newton. In his *Opticks* Newton wrote: "In a very dark Chamber, at a round Hole, about one third Part of an

Inch broad, made in the Shut of a Window, I placed a Glass Prism, whereby the Beam of the Sun's Light, which came in at that Hole, might be refracted upwards toward the opposite Wall of the Chamber, and there form a colour'd Image of the Sun." It was this experiment that led Newton to discover that his observed solar spectrum did not originate in the glass prism but was a property of the sunlight itself. The prism merely refracted the different colors at different angles. This simple arrangement was the first spectroscope, and Newton's experiment marked the first application of spectroscopy to the study of the interaction of light and matter.

It is now common knowledge that the colors observed by Newton correspond to electromagnetic waves of various frequencies, each with a specified wavelength: the higher the frequency, the shorter the wavelength, and vice versa. For example, the frequency of visible light extends from about 4,300 to 10,000 trillion cycles per second, corresponding to wavelengths of about 7,700 to 3,900 angstrom units (ten-billionths of a meter). The angle of refraction of a particular colored ray through a piece of glass is a unique function of its wavelength: the shorter the wavelength, the larger the angle of refraction. Although Newton did not advocate a wave theory of light, he described the various colors by their "degree of refrangibility" as they passed the prism of his spectroscope.

A great deal of useful information is contained in the characteristic spectra associated with the various species and states of matter. Indeed, one might say that matter communicates with us by means of the spectrum of light that it emits and with which it interacts. Consider the visible spectrum of the sun as seen through Newton's prism. The dominant feature of the solar spectrum is a color continuum extending over the entire visible range from red to violet. The distribution of intensity among these colors is governed by the temperature of the emitting surface of the sun. The particular color combination that appears to us as white sunlight indicates a surface temperature of about 6,000 degrees centigrade.

In fact, at a given temperature all hot bodies, regardless of their composition, emit a continuous spectrum of rays with an identical distribution of intensity. As the temperature of the body is increased, this distribution changes: the color being emitted at maximum intensity shifts toward the violet end of the spectrum.

Thus stars that have higher surface temperatures than the sun appear blue against the night sky. Similarly, as the filament of an incandescent electric light bulb is gradually heated, its dominant color changes from an initial dull red to a bright yellowish white.

If Newton had made a very narrow slit instead of a hole in the "shut" of his window, he would probably have discovered that his solar spectrum also contained a sprinkling of narrow, dark lines. A century passed, however, before these delicate features came to the attention of Joseph von Fraunhofer, a German master of optical devices. Fraunhofer mapped hundreds of these dark lines, lettering eight of the most prominent A through H. Another 50 years passed before Fraunhofer's work led to the exciting discovery that many elements found on the earth also exist on the sun. It then became clear that the Fraunhofer lines were caused by the passage of solar rays through the sun's outer atmosphere. Layers of gas in this atmosphere contain isolated atoms of certain elements, which characteristically absorb the sun's rays only at sharp and well-defined wavelengths. For example, a pair of closely spaced dark lines in the yellow region of the solar spectrum— Fraunhofer's D lines—are due to the absorption of sunlight by sodium atoms.

The absorption lines of sodium and other elements can be reproduced in the laboratory by means of a simple absorption spectroscope. When one views an ordinary incandescent lamp through a prism spectroscope with a narrow slit, one sees a continuous spectrum of color extending from red to blue. Now, if one places a flame containing sodium atoms between the lamp and the slit, the continuous spectrum is altered in the yellow region, where the Fraunhofer D lines are located. In fact, the intensity of the light is diminished by an appreciable amount precisely at the wavelengths of the two D lines. This reduction in intensity is caused by the absorption of light by the sodium atoms in the flame. An element that strongly absorbs an incident light ray at a definite wavelength may become entirely transparent at a slightly different wavelength. The width of such an absorption line is defined by the range of wavelengths within which strong absorption takes place.

The existence of characteristic absorption lines is an important aspect of the optical properties of matter in all three of its states: gas, liquid and solid. Isolated atoms or molecules in a gas at moderate pressure yield sharp, narrow absorption lines, which become somewhat broader as the pressure is raised. In liquids and solids these absorption lines become very broad, in some cases encompassing sizable regions of the visible spectrum. Thus red glass examined through an absorption spectroscope shows a strong absorption band covering the green and blue regions and leaving the red region transparent, whereas blue glass shows a strong absorption band in the red and yellow regions and is transparent in the blue region. A completely transparent glass of course shows no absorption bands in the visible region of the spectrum.

The absorption characteristics of materials are not restricted to the visible region of the spectrum. A crystal transparent to visible light may be completely opaque in the infrared and ultraviolet regions. Metals, on the other hand, reflect visible light and hence are opaque in this region, but they are often transparent at short ultraviolet wavelengths.

Simultaneous with the discovery that elements have characteristic absorption lines, it was found that elements are also capable of emitting characteristic radiation at well-defined wavelengths. These emission lines were first observed in flames and later in electrically excited gases. In recent years they have also been produced by electrically or optically excited impurity atoms in certain solids, a situation that is quite similar to that in a gas. When the spectra of such light sources are analyzed in a prism spectroscope, a series of sharp bright lines is obtained. The wavelengths of the emission lines for a particular element coincide exactly with that element's characteristic absorption lines. Sodium, for example, which strongly absorbs at the Fraunhofer D lines, also emits radiation at these wavelengths.

During the 1850's it was further recognized that some of an element's emission lines could be strongly "reabsorbed" by the same element. A related optical property has been added to this list in our own century. Under appropriate conditions matter is capable of amplifying instead of absorbing an incident light ray. This property—the basis of the modern laser—will be discussed later in this chapter.

Toward the end of the 19th century there was collected a vast body of data on the precise wavelengths of the absorption and emission lines in the spectra of a great many elements. Moreover, curious regularities were recognized in some of these spectra. The interpretation of these regularities became a major challenge of the time. The precision with which the spectral lines were charted provided one of the keys to quantum mechanics—the ultimate interpretation of the optical properties of matter.

In the initial formulation of quantum mechanics hydrogen played a decisive role. Its simple atomic structure of one electron bound to one proton produced a line spectrum that revealed the quantum nature of an atom's electronic structure in its barest essentials. These laws were then generalized and applied to the optical properties of more complex atoms containing many electrons and finally to atoms in the liquid and the solid states.

In their present form the laws of quantum mechanics in principle embrace all the optical properties of gases, liquids and solids. The mathematical manipulation of these laws becomes exceedingly intricate, however, when one is dealing with many-body interactions, particularly in solids. For the purposes of this discussion, therefore, the hydrogen atom will serve as a starting point for further generalization about more complex systems.

The hydrogen atom can be described in quantum-mechanical terms as an electron "cloud" surrounding a single proton nucleus, the volume of the cloud being much larger than that of the nucleus. Although the electron behaves in some respects as a pointlike particle with a definite charge and mass, in other respects its position can be regarded as being spread over an extended volume whose size and shape depend on the electron's motion. The density of this cloud at each point around the nucleus represents the probability of finding the electron at that point. The total internal energy of the atom is uniquely determined by the configuration of the electron cloud, and the configuration is in turn governed by the wavelike behavior of the electron.

An atom can exist in only one of a number of quantized energy states, each state corresponding to an electron cloud of a different size or shape. Accordingly an atom can change its energy only in distinct quantized steps, each step a transition from one energy state to another. The various states, arranged in order of increasing energy, constitute the energy-level diagram of the atom.

Atoms that have more than one electron can be similarly described. In that case, however, the different electrons occupy different quantum states, and the internal energy arises from electron-electron interactions as well as from electron-nucleus interactions. Most of the chemical and optical properties of an atom are determined by the quantum states of its valence electrons, which occupy the atom's outer boundary. The rest of the electrons form closed inner shells and are tightly bound to the nucleus. An atom of sodium, for example, has 11 electrons, one of which is the valence electron responsible for the chemical behavior of sodium as an alkali metal. The distinct energies associated with the various quantum states of this single valence electron are responsible for the optical properties of isolated sodium atoms in the gaseous phase.

There are several ways of looking at the system of quantum states in a solid. One of these views—the energy-band model—traces the quantum states of the solid back to their origin in the isolated atom. This is done by assuming that the atoms of a given solid are perfectly arranged on a fictitious crystal lattice with interatomic separations many times larger than those found in an actual solid. The quantized states of this fictitious solid are simply duplicates of the states of an isolated atom that undergoes negligible interactions with its neighbors. As the interatomic separation is gradually reduced toward the value corresponding to that for an actual crystal lattice, however, the energy levels of the atoms split into broad bands [*see the illustration on the facing page*]. This splitting occurs when the valence electron clouds of adjacent atoms begin to overlap appreciably, giving rise to strong interactions among the atoms. The overlapping of the various bands in metals is at the root of both their electrical conductivity and their distinctive optical properties, including such absorption characteristics as color and opacity. On the other hand, in an insulator such as sodium chloride or calcium fluoride the valence electrons occupy nonoverlapping bands and the crystals are generally transparent.

In addition there exist a variety of colored materials that derive their optical properties from the quantum states of impurity atoms embedded in the crystal lattice of an otherwise transparent solid. The width of the energy level of the impurity atom depends on the extent of its interaction with the host lattice. For a strongly interacting state the level is broad and forms a band similar to

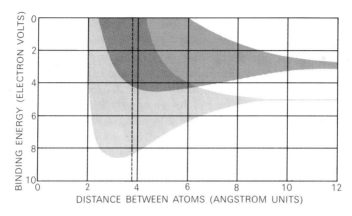

Quantum states of solid sodium can be traced back to their origin as energy levels in the isolated sodium atom. This is done by assuming that the atoms of the solid are arranged on a fictitious crystal lattice with interatomic separations many times larger (right) than those found in the actual solid. As the interatomic separation is gradually reduced toward the value corresponding to that for the actual crystal lattice of sodium (broken vertical line), the energy levels of the atoms split into broad bands (left). The overlapping of the bands in metals is responsible for such distinctive optical properties as color and opacity. Only the 3s and 3p bands of sodium are shown here.

that of a pure solid; otherwise the level is narrow and resembles that of an isolated atom. For example, sapphire is a transparent ionic crystal consisting of aluminum oxide with traces of titanium and chromium, whereas ruby has the identical composition but with a few percent of the chromium. The absorption spectrum of a ruby shows wide absorption bands in the blue region, resulting in the ruby's characteristic pink color. The chromium ions, which substitute for aluminum ions in the crystal lattice of the ruby, are solely responsible for this absorption spectrum. The width of the bands shows that the chromium atom interacts quite strongly with the lattice.

A closely associated substitution can take place in certain ionic crystals such as potassium chloride. In this case missing chlorine atoms in the crystal lattice are replaced by free electrons, creating "color centers" with a characteristic absorption spectrum that gives the normally transparent crystal a purple or blue color depending on the temperature.

When an atom is in its lowest energy state, it prefers to stay there for an indefinite time unless it is disturbed by some external means. Such a disturbance can take the form of a collision with an external electron, causing the atom to be excited suddenly to a higher energy state. Once the atom is in an excited state, it tends to decay spontaneously to a lower energy level. The decay is accompanied by the emission of a light wave at a frequency that is uniquely and universally proportional to the change in the energy of the atom. This universal relation of energy change to frequency defines a resonance frequency for each pair of levels. In fact, whenever a spontaneous transition from a higher to a lower level occurs, an electron in the atom exhibits a decaying oscillatory motion at the resonance frequency of the corresponding pair of levels. The oscillation, in turn, is responsible for the light wave radiated at that frequency. When the emission ends, the emitted energy in the light wave precisely equals the change in energy of the atom. Thus the energy in the emitted light wave is also universally related to the frequency of that wave. Here the quantized nature of the atom goes hand in hand with the quantized state of the radiation field; one photon, or quantum of light, is emitted when one atomic transition occurs. The laws of quantum mechanics make it possible to calculate exactly the probability of spontaneous emission from one level to a lower one. The probability is high for some pairs of levels and exceedingly low for others.

In brief, the quantum states of an atom are defined by a set of energy levels, with a resonance frequency associated with each pair. An atom can decay with a predictable probability from a higher to a lower level by spontaneously emitting a photon at the corresponding frequency. It must be emphasized that the exact energies of an atom's quantized states are predictable theoretically; one merely needs to know the number of electrons in the atom and the nature of its nucleus. The rest follows from the universal laws of the interactions of the electrons with the nucleus and with other electrons.

The emission line spectra of isolated atoms can now be interpreted in terms of this quantum-mechanical picture. For instance, the energy-level diagram of a sodium atom contains two closely spaced levels immediately above its lowest energy state. The

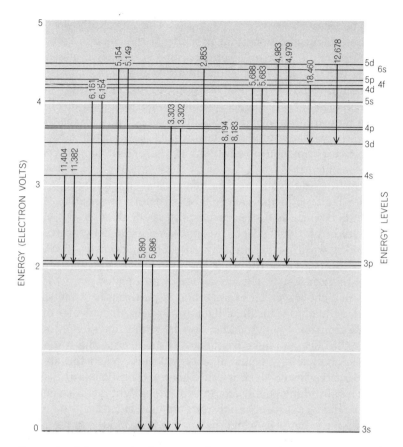

Energy-level diagram of sodium atom shows some of the upper energy levels to which the valence electron (normally at the 3s level) can be excited by the input of energy to the atom. An excited electron can return to the 3s level by a variety of routes (arrows); for each transition from an upper level to a lower one the atom emits a photon of light with a characteristic wavelength, which is indicated in angstrom units on the arrow representing that transition. The two transitions from the split 3p level to the 3s level are responsible for the characteristic yellow D lines in the spectrum of sodium.

spontaneous transitions from these two levels to the lowest energy state are responsible for the emissions at the two closely spaced Fraunhofer D lines. Precise measurements of the difference in wavelength between these two lines reflect the difference in energy between the two excited levels. This "level-splitting" is a manifestation of an important property of the electron known as spin. In a crude sense an electron can be visualized as a spinning entity with a fixed angular momentum. In the more sophisticated theory the electron spin and its resulting interaction are shown to be necessary consequences of the laws of quantum mechanics when they are formulated in a manner consistent with the theory of relativity. These fundamental insights have all emerged from delicate experimental observations of the optical line spectra of the elements.

The structural detail of an atomic nucleus introduces additional minute features in the energy-level structure of an atom; these generally appear as further level-splittings or small energy shifts. The study of these effects in optical emission-line spectra has yielded a wealth of information about atomic nuclei, including their size, charge distribution and spinning behavior.

Let us turn now to the interaction of an isolated atom with an incident light wave. This interaction is particularly strong if the wavelength of the incident light is close to that of one of the atom's emission lines. In other words, strong interaction occurs when the incident light frequency is near or at the resonance frequency of a given pair of energy levels. When the atom is found in the lower level, it undergoes a transition to the upper level by absorbing energy from the incident light ray. The amount of absorbed energy is exactly the amount gained by the atom. The opposite happens when the atom is found in the upper level: the atom decays to the lower level by giving its energy change to the incident light wave. The latter process is an emission act induced by the applied field and drastically different from the spontaneous emission described above. In the induced emission the emitted light wave cannot be distinguished from the incident light. Spontaneous emission, however, is independent of an incident light ray and generally occurs isotropically—the emission probability is the same in any direction in space.

The induced emission probability, which is proportional to the intensity of the incident light, is exactly identical with the probability of the absorption process of an atom initially found in the lower level. An induced emission, however, must compete with the act of spontaneous emission, and at a low intensity spontaneous emission may predominate. Finally, the induced emission (or absorption) probability diminishes considerably if the frequency of the incident radiation is appreciably different from the resonance frequency of the atomic transition.

Induced emission by atoms in the upper level enhances the incident light, whereas absorption by atoms in the lower level attenuates it. Accordingly when the average number of atoms in the lower level exceeds that in the upper one, the absorbing transitions would prevail and a total attenuation of the light wave would result. The reverse process, called population inversion, occurs if the average number of atoms in the upper level is larger than the number in the lower—the incident light is then amplified as it passes through the medium. This kind of amplification, "light amplification by stimulated emission of radiation," is the underlying principle of the laser.

There exist a host of nonradiative processes that also cause transitions among atomic energy levels. These play an important role in determining the average number of atoms in each level and therefore establish the main absorption and emission characteristics of an ensemble of atoms. For instance, the impact of a free electron with an atom or the collision of two or more atoms can induce transitions among various atomic energy states without requiring either the emission or the absorption of a light wave. In such cases the quantized change in the internal energies of the atoms is determined by the kinetic energies of the collision partners.

For matter in thermal equilibrium the exact number of atoms in each level is uniquely determined by the temperature of the system. It is a general property of this thermal distribution that the average number of atoms in a lower level is always larger than that in an upper one. Because of this, matter in thermal equilibrium always attenuates an incident radiation at a frequency on or very near any atomic resonance frequency.

A broad class of systems exist that, though in a steady state, are not in thermal equilibrium and therefore cannot be described in terms of temperature. Most of the universe is composed of such nonthermal systems. In these cases the average number of atoms may be larger in a lower or an upper level, depending on their excitations and decays. These systems allow amplification of an incident radiation when an upper level is more populated than the lower one.

Consider an ordinary neon lamp, essentially a glass tube filled with neon gas at low pressure. An electric voltage applied across a pair of electrodes at opposite ends of the tube causes the ionization of a small fraction of the neon atoms. This results in a stream of free electrons—an electric current—passing through the tube. The electrons collide with neon atoms and excite them into higher energy levels. The spontaneous emission from the various excited levels is responsible for the color of the lamp.

In such an electrically excited gaseous discharge the atomic population distribution is completely nonthermal. Under the proper conditions it could allow population inversion among some level pairs and amplification of light waves at the corresponding resonance frequencies. It was an oversight in scientific history that this possibility passed unrecognized in the late 1920's, since by that time quantum mechanics had successfully formulated the principles of induced emission and absorption of light, and the excitation processes in electrically excited gas discharges were also fairly well explored. All the necessary theoretical and technical information for the achievement of light amplification in an electrically excited gas was close at hand. However, physicists of the time were so preoccupied with the emission and absorption line spectra and the characteristics of matter in thermal equilibrium that they missed this exciting possibility. Thus laser amplification, based on the principle of population inversion in a gas discharge, was achieved some 30 years late.

A medium capable of a sufficient light amplification becomes unstable if it is placed between a pair of parallel mirrors. The light wave, propagating at right angles to the mirrors, will be reflected back and forth and enormously amplified by the medium. This process can be initially triggered by the spontaneous emission of a light ray in the amplifying medium itself. Output coupling can be provided by allowing a small beam of light to be transmitted through the surface of a mirror. This system, commonly called the laser, generates an intense directional light with a minute frequency spread. Several general types of laser exist, each of them using a different method to prepare level pairs with inverted populations. Among the major types are gas-discharge lasers, chemical lasers, semiconducting-diode lasers and homogeneous optically pumped liquid and solid-state lasers. These devices, useful in many practical applications, have also provided a new tool for research into the nature of matter.

Consider now an atom in its lowest energy level interacting weakly with a light ray whose energy does not correspond to the separation of any upper level from the ground state. It is no longer appropriate to describe this interaction in terms of absorption and emissions accompanying a real atomic transition, since the frequency of the light is appreciably different from the resonances of atomic transitions. Actually the atom serves to scatter the incident light weakly in all directions while remaining essentially in its unperturbed energy state. The incident radiation will force an atomic electron to oscillate at the frequency of the light wave itself; the electron, in turn, radiates the scattered light wave in all directions at its oscillating frequency. The intensity of the reemitted light is generally a function of the incident light frequency and the proximity of this frequency to various atomic resonances.

When a collection of atoms is subjected to such an incident light wave, the atoms' reemitted waves interfere with one another. The propagation pattern of the resultant sum of the reradiated light waves depends on the density and spatial distribution of the scattering atoms. This effect is responsible for the index of refraction of a transparent material, giving rise to the reflection and refraction of light. The reflection of an incident light by a transparent solid with a uniform density comes from the superposition of reemitted light waves by atoms at the surface boundary of the medium. The refracted light ray results from this superposition of reemitted light by atoms in the bulk of the solid.

In metals the reradiation at the boundary interface is so strong that most of the incident radiation is reflected over a wide range of the spectrum. This phenomenon differs, however, from the partial reflection of light at the boundary of a transparent dielectric crystal. The conduction electrons in metal behave collectively as a highly dense plasma, and cause total reflection of long-wavelength incident light; at shorter wavelengths there is partial transmission. Gold, for example, reflects red and yellow light strongly but allows some penetration by green rays, which are then completely absorbed within a small thickness of the bulk of the metal. Silver, on the other hand, reflects strongly over most of the visible spectrum but allows considerable transmission in the ultraviolet.

There are in addition a number of light-scattering processes in which an incident ray at one frequency is absorbed by an atom and then reemitted at an entirely different frequency. When the absorption and reemission occur simultaneously as part of a single transition, the phenomenon is called the Raman effect. Here the difference in frequency between the two light waves must exactly equal a resonance frequency between a pair of atomic levels; the individual light frequencies, however, are not required to be near any of the atomic resonances. Raman scattering, which can take place in gases, liquids or solids, is usually weak unless the intensity of the incident light is very high.

Light absorption at one frequency and light emission at another frequency can also take place in two separate steps. An incident ray at the resonance frequency of a pair of levels is first absorbed in the usual way, causing an atomic transition from a lower level to an upper one. After a certain delay the atom spontaneously undergoes a transition involving a third level, emitting a light ray at the appropriate frequency. In some cases the delay may be as short as a microsecond or less, in which case the phenomenon is called fluorescence. In other cases the delay can be for seconds or even days, allowing the emitted light to be observed after the incident light is turned off; this phenomenon is called phosphorescence.

The absorption and emission characteristics of matter outlined in this chapter of course extend far beyond the visible region of the electromagnetic spectrum. On the long-wavelength side there are the infrared, far-infrared, microwave and radio-wave regions, and on the short-wavelength side the ultraviolet, X-ray and gamma-ray regions. The optical properties of materials in all these regions are similar to those in the visible region in their basic relation to the quantum-mechanical interactions of electrons, but they are vastly different in detail. In these ranges spectroscopy requires a diversity of experimental techniques, ranging from Newton's prism to radio-wave and gamma-ray spectrometers.

Now that the properties of all materials are better understood, it is clear that quite different materials can be used for the same purpose. This calls for subtle choices involving both technology and economics.

W. O. ALEXANDER

The competition of materials

The engineer necessarily takes a greater interest in the properties than in the composition of the materials he chooses to meet the requirements of his design. Nowadays he is likely to find at least two and in some cases three or four radically different kinds of material ready to serve in each end use. In such industries as aerospace and electronics, where cost has not been a ruling consideration, designers have produced numerous dramatic demonstrations of the interchangeability of materials that is a recurrent theme of the preceding chapters of this book. This trend in technology has gained economic significance in western Europe and the U.S. over the past decade as producers in one primary materials industry after another have developed capacity in excess of the demand for their product. Now producers as well as users are promoting the competition of metals, ceramics, glasses, plastics, rubbers, concrete and timber, each defending its traditional markets as it seeks entry into the markets of others.

Here is a set of new and knotty questions for the business economist. The plotting of supply-demand curves for a given commodity no longer suffices to explain or predict its movement.

Simple comparison of the cost per pound of steel, aluminum and plastics tells little about their true relative competitive positions. Economics must reckon at closer range with technology in order to estimate the viability of one material against other materials. The detailed investigations necessary to establish the ruling criteria of the many-cornered rivalries of materials technology remain to be made. The preliminary considerations presented in this chapter suggest, however, that a useful index of the competitiveness of a material is its cost per unit of property required.

Cost in this formulation is the cost of the material in the job. To the cost of the material at the mill, plant or warehouse is added the cost of fabrication and incorporation in the final product. As for properties, those that determine the usage of the vast bulk of materials can be listed as strength and rigidity, space-filling and the quality and durability of the surface. The last requirement may be met by a number of accessory materials or treatments, so that it is not usually a critical element in the choice of the primary material. The space-filling requirement can be met in many applications by leaving the member hollow, so that this is not critical either, unless great rigidity is required. Strength remains as the property most sought. The index of competitiveness may therefore be expressed as cost per unit of strength.

There are, of course, applications where high electrical conductivity or extreme rigidity or high corrosion resistance may be of transcendent importance. Such requirements, however, dominate the selection of only a minor portion of all the materials that are used commercially. The principal exception is copper, for which high conductivity dictates more than half the usage.

The national accounts of the industrial countries customarily reckon the costs of materials at three stages on their way to market. The first stage is extraction: the cost of ores at the mine (which may include a charge for beneficiation) or the cost of timber at the forest (which in the future must include the cost of afforestation). The second stage is at the conversion of the raw material into purified or concentrated form, for example a metal in the pig or ingot. Finally there is the cost at which, in the case of metals, the plate, sheet or shape is delivered to the manufacturer of the end product. At each stage cost is incurred for the energy involved in the conversion or processing, for labor and for selling, administration and other costs, all of which go to build up the final price.

In the production of metals some credit can be taken for the reprocessing of scrap, offset in part by the loss in value of worked metal that must be scrapped and recycled in the mill. The production of plastics begins at the oil or gas well. To the fabricator a plastic is usually delivered in the monomer state, to be polymerized in the same operation that yields the end product. Plastics typically yield a low order of scrap, usable only in such secondary products as toys, that is a burden to the solid-waste disposal system. The felling of timber and dressing of lumber produce waste in vast quantities, useful for by-products such as chipboard and fiberboard but so cheap that its utilization is usually a question of transportation costs. Cement and concrete involve the simplest production cycle: the calcining of limestone rich in silica, alumina and iron oxide, and grinding the sinter with a small amount of gypsum. This is followed by mixing the cement with aggregate and sand, usually on the site, and pouring the slurry into molds that contain the steel reinforcing bars. Concrete has little or no value at the end of service.

In the past a producer of one of these commodities could gauge his efficiency merely by keeping his eye on other producers of the same commodity. Now he must operate against the unfamiliar economics of entirely different kinds of materials. It is important for the steel producer, with costs at the ingot or raw steel running at $85 per ton, to know that the pig cost of the aluminum producers is around $500 per ton. Any substantial reduction in the cost of aluminum pig is bound to make this metal a more formidable competitor to steel. The costs at the purified metal stage reflect the large differences in the geology and chemistry of the various metals and the corresponding differences in the costs of mining and smelting them. Costs thereafter tend to converge because the operations involved—rolling, forging, casting and so on—are more or less the same for all. Over the past 25 years, as producers of aluminum, magnesium, beryllium, titanium and the other new or relatively new metals have perfected their primary and secondary technologies, costs of purified metals have been the determining factor in the development of markets for these metals.

The general price trend for most materials has been upward. In the U.S. the slope is about the same as that of the economic inflation. In western Europe metal prices have risen somewhat more slowly since the end of World War II, reflecting efficiencies

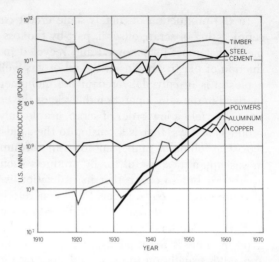

Competition of six major materials in U.S. is reflected in curves of poundage that are based on similar ones plotted by C. G. Suits of the General Electric Company.

gained in new plant. The marked exception to the rule on both sides of the Atlantic has been the prices of the major plastics; these are still decreasing, and plastics are therefore becoming correspondingly more competitive.

Statistics for output divide the principal materials into two classes. In the U.S. the older materials show a rather slow increase [see chart above]. Aluminum and the plastics, in contrast, rise steeply from low levels of output, increasing at 10 percent and 10 to 15 percent per annum respectively. The same trends, trailing the U.S. by two to five years, appear in the western Europe statistics. At about five million tons in 1962, the U.S. plastics output might seem negligible compared with that of steel, at about 100 million tons. An important property of plastics, however, is their low density. The cubic-foot curve plotted by C. G. Suits of the General Electric Company, shows the plastics beginning to bulk at the same order of magnitude as steel in the U.S. economy [see chart on facing page]. Even bigger than steel, and by a huge margin, is concrete. The tonnage for any year can be derived by multiplying cement output by 6; for volume, the multiplier is 8.

The retardation in the continued growth of output curves for the older metals undoubtedly reflects the competitive inroads

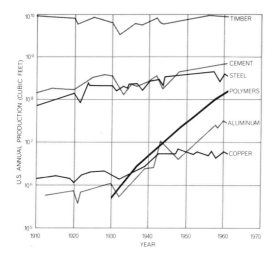

Relation of curves is changed by using cubic feet instead of pounds, correcting for lighter weight of newer materials. On this basis polymers begin to approach steel.

made by aluminum and the plastics. It also reflects increasing efficiency in the engineering of metals and in some cases a disproportionately lower use of them in major applications. For the generation of electric power, which is increasing 8 to 10 percent per annum, the use of copper in generators has been reduced from 200 pounds per megawatt to 56 pounds per megawatt during the past decade. The increase in speed and payload from the Boeing Stratocruiser to the Boeing 707 has reduced the use of metal from 7.7 to three pounds per passenger mile. To a less spectacular degree, increase in the efficiency of design has reduced the metals requirements of the automobile industry and the heavy construction industry, particularly in factory buildings and bridges.

A summary economic expression of these developments in technology is provided by the technique of input-output analysis developed by Wassily Leontief of Harvard University. Putting the input-output tables of the U.S. economy for the years 1947 and 1958 to the task of producing (in the computer) the same 1958 gross national product, Leontief's colleague Anne P. Carter found a net reduction of 10 percent in the materials inputs required by the 1958 technological order. The metals showed the biggest losses in usage, with a 27 percent drop in the requirement for steel [see

"The Economics of Technological Change," by Anne P. Carter; SCIENTIFIC AMERICAN, April, 1966. Offprint 629.]

It appears, therefore, that the industrial nations have arrived at a significant turning point; their per capita consumption of metals has reached a peak and is now on the decline. Before one applies this finding to a projection of the future growth of the metals industries in the world as a whole, however, it is well to consider one or two contingencies. The per capita consumption of metals for two-thirds of the world population has yet to approach that of western Europe, America and Japan. Should money and technical staff be redeployed to increase the rate of development of the poor nations, the present spare capacity in the industrial nations would be quickly overfilled. One has only to think of the 760 million people in China, who consume 90 pounds of steel each year compared with the half-ton consumed per capita in the U.S., to appreciate that a mere trebling of the rate of consumption in China would soak up most of the world's unused steel-producing capacity. If the people of India and China were to open one can per week, that would be some 60 billion cans, or about two million tons of steel strip, a year.

For the present, excess capacity in the metals industries sharpens the competitive struggle. One significant conflict engages modes of fabrication rather than kinds of metals. In the metal trades it has been axiomatic that the properties and dimensions of metals in the wrought form are universally superior to those of metals in castings. Inaccuracy and uncertain reproducibility in the making of the molds and poor surface appearance and low mechanical properties in the castings have nullified the economies inherent in moving directly from the molten metal to the finished product. In recent years, however, technology has been moving faster in the foundry than in the metal-stamping and forging shops. Such techniques as vacuum-pressure diecasting and the automatizing of investment casting have been reducing costs and improving the product. Whereas wrought steel continues to come in at lowest cost, casting brings in products made of nonferrous metals at a lower cost than products made of these metals in the wrought condition. In the United Kingdom the cost-per-unit-strength criterion finds wrought steel at .05 to .13 pence per ton of tensile strength per inch compared with aluminum castings at .18 to .45 pence and wrought aluminum at .20 to .31 pence.

Engineers and metallurgists in the wrought-metal field have missed significant opportunities for progress as a result of their predilection to go automatically for larger and larger initial ingots to feed bigger and bigger breakdown mills, presses and forges. An elementary analysis of the cost of working metal by rolling shows that a great amount of energy is required to break the ingot down to slab or billet and then roll it to finished dimensions. Energy costs are bound to become dominant with continued increase in the price of energy and rising production per man-hour. Already energy accounts for about a third of the total cost of converting raw metal to finished intermediate products, with direct labor and overhead each accounting for a third. Although gigantism secures some marginal economies, it can be justified in the future only in the rolling of wide plate and sheet and in the forging of heavy single pieces. The alternative approach, just gathering momentum, is continuous casting. For small cross sections of steel and aluminum, continuous casting has already demonstrated its feasibility and economy. In addition to reducing energy charges, the technique involves smaller plant costs and greater freedom in plant location, and its high speed makes it possible to keep inventories to a minimum.

The heavy construction industry presents an arena in which steel is pitted against concrete. At present the situation in this tug-of-war varies from country to country. It appears to depend less on the objective qualities of the two materials than on the ingenuity and prestige of the architects and engineers committed to the one material or the other and on the competitiveness and marketing flair of the producers in the two industries. Judging by North America, it would seem that steel has won in vertical structures whereas concrete has taken over horizontal structures, except for the very biggest bridges. In the United Kingdom the competitive position of steel is compromised by building codes that recognize only 45 percent of the 100 percent improvement in the strength of structural steel achieved since 1877. Regulations require, in addition, heavy fireproofing of steel structural members, a measure usually secured by cladding the member in concrete with a consequent considerable increase in the dead load of the structure. Americans minimize this requirement by insisting on a low fire rating for the entire contents of the building.

On the other hand, the proponents of steel in both Britain and America must contend with a 600 percent improvement in the performance of reinforced concrete achieved in recent years. On the cost-per-unit-strength basis structural steel appears evenly matched with reinforced concrete. It is no wonder that relatively small multistory buildings can be built of concrete at costs 25 to 35 percent below that of steel.

On its lighter flank, steel has been losing other markets to aluminum. Although the output of aluminum in the U.S. economy comes to only 5 percent of the tonnage of steel, it should be noted that this amounts to 15 percent by volume. Moreover, 50 percent of the aluminum goes into identified steel markets. Aluminum made an easy entry into many of these markets during the war and in the immediate postwar years, when steel was in short supply. But aluminum has continued its encroachment and competition has sharpened, particularly in the U.S., as both industries have gone out to sell excess capacity. On the basis of cost per unit strength, experience in the United Kingdom places steel well ahead of aluminum, at .05 to .13 pence compared with .20 to .31 pence per ton of tensile strength per inch. Secondary criteria, such as lightness, rigidity and better corrosion resistance, must therefore enter into economic design considerations to qualify the rule of the primary criterion.

In the U.S. the competition of aluminum has induced major steel producers to establish research and product-engineering laboratories that they never seemed to require for competition with one another. Toward recovery of the substantial portion of the container industry lost to the light weight, easy formability and corrosion resistance of aluminum, they have reduced by two-thirds the thickness of strip rolled for cans and are looking for ways to eliminate the expensive tin and tinning process. In the construction industry, steelmakers turned back a challenge from aluminum alloys for short-span highway bridges, but they had to share this victory with reinforced concrete. Conversely, it appears that steel has lost the skyscraper curtain-wall market. Here the properties desired are lightness, appearance and constancy of surface. Aluminum, anodized to various colors and finishes, has proved more desirable than even stainless steel and much less expensive.

In the shipyards of the United Kingdom, on all new passenger ships and on the vast majority of cargo vessels, the use of aluminum alloys is reducing the weight of top hamper—deckhouses, lifeboats and the like—and hence the weight of metal per ton mile of cargo moved. This use of aluminum also brings improvement in the stability of the ship.

Aluminum has grown at the expense of another traditional metal: copper. The latter has now been displaced from long-distance electrical transmission lines. What aluminum lacks in conductivity it makes up in cheapness and lightness. Use of the metal as a specially shaped rod, rather than in a twisted wire cable, produces a conductor with much greater efficiency on the criteria of cost and weight per kilowatt of power transmitted. Aluminum holds its advantage down to 11 kilowatts and even lower, leaving copper to be used in central- and substation gear.

In every one of their markets the metals producers encounter competition from plastics. What makes this competition difficult is that no catalogue of the costs and properties of the various plastics indicates very clearly what their advantages are. On the basis of cost per unit strength, the most competitive of these materials—polyvinyl chloride and resin and glass fiber laminates—show the comparatively high values of .9 to 8.0 pence respectively. Polyethylene, which has now swept to first place among plastics, costs in molded form as much as .9 pence per unit strength. The answer is, of course, that the vast bulk of plastic materials is not now purchased for performance on this criterion. On the contrary, their great attractions are falling price, low density, the facility with which they can be worked into intricate finished shapes, and the wide range of hardness and softness, stiffness and flexibility, color, transparency and opacity in which they can be specified. When all is said and done, however, the position of plastics in the materials marketplace can be epitomized by the view of a plastic engineer I know, who says he always likes to sell plastics but he would prefer to buy metals any day.

It is not surprising that plastics should dominate the toy business or should now be on the ascendant in the packaging industry. From copper, lead, zinc, tin, steel and cast iron, however, polyethylene and polyvinyl chloride have taken large volumes of pipe and tube for domestic, municipal and industrial plumbing. Glass-filled nylon is used instead of zinc, magnesium or aluminum in precision die castings. In the U.S. resin-glass laminates have made the manufacture of small boats into a big business. These are reasonably high-specification products, and the success of plastics in such uses is a portent of further developments. In the case of metals, limits and properties are now well defined and no significant new metals are in the offing. In plastics, where new molecules remain to be synthesized, polymerized into chains of new configuration and aggregated into structures with novel combinations of properties, there is always a chance of an overnight revolution. As Herman F. Mark shows [see "The Nature of Polymeric Materials," page 85], if a plastic could be developed with suitable mechanical properties at 150 degrees centigrade, this could completely alter the competitive situation in such huge metal markets as automobiles and construction. This is the biggest development impending in the technology and economics of materials.

The authors

W. O. Alexander
("The Competition of Materials") is professor of metallurgy and head of the department of metallurgy at the University of Aston in Birmingham, England. He is a graduate of the University of Birmingham, where he studied metallurgy from 1928 to 1937. From 1937 to 1961 he was engaged in metallurgical work at Imperial Chemical Industries. He writes: "So far as my leisure-time pursuits go, I suppose I can be considered first of all a sportsman since I played Rugby football for some 15 years and in my early days played a lot of tennis and in more recent years have swum regularly." He is also interested in the organization of various types of amateur sport. These activities, he says, "only leave me time to take a reasonably intelligent interest in drama and painting and whenever possible the wide-open spaces." The principal content of Alexander's article was originally presented in *Contemporary Physics;* the permission of that journal to repeat it here is gratefully acknowledged.

R. J. Charles
("The Nature of Glasses") is at the General Electric Research and Development Center in Schenectady. He writes: "Originally a Canadian, I was attracted by the open life of the mining engineer and took a B.S. in mining in 1948 and an M.S. in metallurgy in 1949, both at the University at British Columbia. Astonished by the total disaster of a gold-mining venture I subsequently embarked on, I returned to my long-standing interest in metals and minerals." He did further graduate work at the Massachusetts Institute of Technology, where he received a

doctorate in 1954 and served on the faculty until joining General Electric in 1956. He writes: "By personal choice my research interests are divided between a long-term interest in theoretical aspects of silicate and oxide systems and shorter-term pursuits selected from an overflowing reservoir of industrial scientific problems." Charles, now a U.S. citizen, says his major outside interest is "the experimental and computer analysis of two-fluid locomotion (sailing)."

A. H. Cottrell

("The Nature of Metals") is Chief Scientific Advisor (Studies) in the British Ministry of Defence. He writes: "Most of my work on the theory of dislocations in metals was done at the University of Birmingham, of which I am a graduate and where I was professor of physical metallurgy. After 1955, when I moved to the Atomic Energy Research Establishment at Harwell, my interests turned toward problems of nuclear radiation damage in solids and later to the theory of fracture, a subject that I continued when I moved to the University of Cambridge in 1958 as the Goldsmiths' Professor of Metallurgy. I have become increasingly interested in recent years in the role of science in national affairs." That interest has led Cottrell successively to membership on the United Kingdom Atomic Energy Authority and the Advisory Council for Scientific Policy, to the Ministry of Defence and to membership in the recently formed Central Advisory Council for Science and Technology.

Henry Ehrenreich

("The Electrical Properties of Materials") is Gordon McKay Professor of Applied Physics at Harvard University. Born in Germany, he came to the U.S. in 1940 and did all his college work at Cornell University, from which he received a Ph.D. in 1955. From then until 1963 he was at the General Electric Research Laboratory in Schenectady. His research activities have involved electron systems, transport properties of semiconductors, optical properties of solids, and the band structure and magnetic properties of noble and transition metals.

John J. Gilman

("The Nature of Ceramics") is professor of physics and metallurgy at the University of Illinois. He took bachelor's and master's degrees at the Illinois Institute of Technology in 1946 and 1948 respectively and received a Ph.D. from Columbia University in 1951. He worked in industry until 1960 and then spent three years as professor of engineering at Brown University before taking up his present work. His research centers on the mechanical behavior of solids; he writes that he "had the good fortune to be the first (together with my then colleague W. G. Johnston) to devise methods for measuring the velocities of dislocations in crystals."

Ali Javan

("The Optical Properties of Materials") is professor of physics at the Massachusetts Institute of Technology. A native of Iran, now a U.S. citizen, he recieved a Ph.D. in physics from Columbia University in 1954 and remained at Columbia for five years as a research associate. From 1958 to 1961, when he went to M.I.T., he was a member of the technical staff of the Bell Telephone Laboratories.

Frederic Keffer

("The Magnetic Properties of Materials") is professor of physics and chairman of the department of physics at the University of Pittsburgh. He did his undergraduate work at the State College of Washington, which is now Washington State University; he writes that "between my desultory senior year in 1939-1940 and the award by the college of a B.S. in 1945 I served for five years in the infantry, touring Europe from Brittany to Leipzig with the Sixth Armored Divison." Keffler obtained a Ph.D. from the University of California at Berkeley in 1952 and since then has been at Pittsburgh. His research, he says, "has centered on the theory of magnetism and has included studies of microwave resonance absorption by ferromagnets, ferrimagnets and antiferromagents; relaxation processes; magnetocrystalline anisotropy; the nature of the short-range coupling force, and types of magnetic ordering."

Anthony Kelly

("The Nature of Composite Materials") has been a lecturer in metallurgy at the University of Cambridge for the past eight years; in October he will join the National Physical Laboratory as superintendent of a new division concerned with inorganic and metallic materials. Kelly was graduated from the University of Reading in 1949 and obtained a Ph.D. from Cambridge in 1953. Although most of his work has been done in Britain, he has had several associations in the U.S. He was at the University of Illinois from 1953 to 1955 and at Northwestern University from 1956 to 1959; earlier this year he was for a time at the Carnegie Institute of Technology. He often spends summers working in industrial and government laboratories; he says he thinks that "is an ideal way to learn what applied science means."

Herman F. Mark

("The Nature of Polymeric Materials") is director emeritus of the Polymer Research Institute at the Polytechnic Institute of Brooklyn. He has long been a leading figure in the field of polymer chemistry, which he entered more than 40 years ago at the Kaiser Wilhelm Institute for Fiber Chemistry in Germany. Mark was born in Vienna and distinguished himself as an athlete as well as a scholar; he played in the Austrian national soccer league as a young man and was also an excellent skier. In 1921 he received a Ph.D. *summa cum laude* from the University of Vienna. He taught at the University of Berlin for a year before going to the Kaiser Wilhelm Institute; in 1926 he joined the research laboratories of I. G. Farbenindustrie, where he worked on studies of cellulose and other polymeric materials. He resigned from the firm when the Nazis took power; leaving Germany, he became director of the First Chemical Institute of the University of Vienna. When the Nazis invaded Austria, he left Europe and worked for two years with a Canadian pulp company before going to the Polytechnic Institute of Brooklyn. Mark wishes to acknowledge the collaboration on his article of his colleague and successor as director of the Polymer Research Institute, Murray Goodman.

Sir Nevill Mott

("The Solid State") is Cavendish Professor of Experimental Physics at the University of Cambridge and a Fellow of the Royal Society. A Cambridge graduate, he began his academic career as a lecturer at the University of Manchester in 1929. From 1930 to 1933, after a period of work with Niels Bohr, he was a lecturer at Cambridge; for 21 years after that he was professor at the University of Bristol. His appointment as Cavendish Professor at Cambridge took place in 1954. In addition to his work at the Cavendish Laboratory he served from 1959 to 1966 as Master of Gonville and Caius College at Cambridge. Sir Nevill writes: "My research interests were in nuclear physics before I went to Bristol and since then they have been in solid state, particularly in the physics of metals and semiconductors and in the theory of photographic emulsions. My present research interest is in electrical conduction in noncrystalline solids, a subject that is rapidly opening up." Sir Nevill has served on various committees concerned with science education and is now chairman of a committee set up by the Nuffield Foundation to frame new methods of teaching physics and chemistry to students between the ages of 15 and 18. He adds: "I am also interested in disarmament and strategic studies and have been present at some of the Pugwash conferences on science and world affairs."

Howard Reiss

("The Chemical Properties of Materials") is director of the North American Aviation Science Center in Thousand Oaks, Calif., and a vice-president of North American Aviation, Inc. After being graduated from New York University in 1943 he worked for two years in the Manhattan project. He then began graduate work in chemistry at Columbia University, from which he received a Ph.D. in 1949. Among his associations since then are Boston University, where he was a member of the faculty for two years; the Bell Telephone Laboratories, where he was a member of the technical staff from 1952 to 1960, and Atomics International, a division of North American Aviation, Inc., where he was successively associate director and director of the research division before taking his present position in 1962.

Cyril Stanley Smith

("Materials") is Institute Professor at the Massachusetts Institute of Technology, attached to both the department of metallurgy and the department of humanities. Born and educated in Birmingham, England, he came to the U.S. in 1924 as a graduate student and received a doctoral degree from M.I.T. in 1926. From 1927 to 1942 he was research metallurgist with the American Brass Company. In 1946, after three years at the Los Alamos Scientific Laboratory, he went to the University of Chicago to establish the Institute for the Study of Metals, which was the first academic laboratory for interdisciplinary research on materials in the U.S. He took his present position in 1961. Smith was an original member of the General Advisory Committee of the Atomic Energy Commission and has served on the President's Science Advisory Committee. He writes that his "main interests are structure, of all things at all levels, and the history of technology and science."

John Ziman

("The Thermal Properties of Materials") is professor of theoretical physics at the University of Bristol. Although he was born in England, he lived for many years in New Zealand and received much of his education there. Returning to England for graduate study, he obtained a doctorate from the University of Oxford. In 1954 he became a lecturer at the Cavendish Laboratory of the University of Cambridge. Since going to the University of Bristol three years ago he has been concerned mainly with the theory of the electronic structure of solid and liquid metals. Ziman was recently elected a Fellow of the Royal Society.

Bibliography

Materials

ANCIENT EGYPTIAN MATERIALS AND INDUSTRIES. A. Lucas and J. R. Harris. Edward Arnold Ltd., 1962.

CHINA AT WORK. Rudolf P. Hommel. The John Day Company, 1937.

DE RE METALLICA. Georgius Agricola. Translated by Herbert Clark Hoover and Lou Henry Hoover. Dover Publications, Inc., 1950.

A HISTORY OF METALS. Leslie Aitchison. Interscience Publishers, Inc., 1960.

A HISTORY OF TECHNOLOGY. Edited by Charles Singer, E. J. Holmyard and A. R. Hall. Oxford University Press, 1954.

LAZARUS ERCKER'S TREATISE ON ORES AND ASSAYING: TRANSLATED FROM THE GERMAN EDITION OF 1580. Anneliese Grunhaldt Sisco and Cyril Stanley Smith. The University of Chicago Press, 1951.

MATERIALS AND THE DEVELOPMENT OF CIVILIZATION AND SCIENCE. Cyril Stanley Smith in *Science,* Vol. 148, No. 3672, pages 908-917; May 14, 1965.

METALLURGY IN ARCHAEOLOGY. R. F. Tylecote, Edward Arnold Ltd., 1962.

NOTES ON PREHISTORIC AND EARLY IRON IN THE OLD WORLD. H. H. Coughlan in *Occasional Papers on Technology: 8.* Oxford University Press, 1956.

ORIGINS OF THE SCIENCE OF CRYSTALS. John G. Burke. University of California Press, 1966.

THE PIROTECHNIA OF VANNOCCIO BIRINGUCCIO. Translated by Cyril Stanley Smith and Martha Teach Gnudi. The American Institute of Mining and Metallurgical Engineers, 1942.

THE PREHISTORY OF SOLID-STATE PHYSICS. C. S. Smith in *Physics Today*, Vol. 18, No. 12, pages 18-30; December, 1965.

The solid state

ATOMIC STRUCTURE AND THE STRENGTH OF METALS. N. F. Mott. Pergamon Press, 1956.

ELEMENTS OF MATERIALS SCIENCE. L. H. Van Vlack. Addison-Wesley Publishing Company, Inc., 1966.

INTRODUCTION TO SOLID STATE PHYSICS. Charles Kittel. John Wiley & Sons, Inc., 1956.

PRINCIPLES OF THE THEORY OF SOLIDS. J. M. Ziman, Cambridge University Press, 1964.

SEVEN SOLID STATES. Walter J. Moore. W. A. Benjamin, Inc., 1967.

The nature of metals

PHYSICAL METALLURGY. Edited by R. W. Cahn. John Wiley & Sons, Inc., 1965.

THE STRUCTURE OF METALS AND ALLOYS. W. Hume-Rothery and G. V. Raynor. The Institute of Metals, London, 1962.

STRUCTURE OF METALS: CRYSTALLOGRAPHIC METHODS, PRINCIPLES, AND DATA. Charles S. Barrett and T. B. Massalski. McGraw-Hill Book Company, 1966.

THEORY OF CRYSTAL DISLOCATIONS. F. R. N. Nabarro. Oxford University Press, in press.

The nature of ceramics

CERAMICS: STONE AGE TO SPACE AGE. Lane Mitchell. McGraw-Hill Book Company, Inc., 1963.

INTRODUCTION TO CERAMICS. W. D. Kingery. John Wiley & Sons, Inc., 1960.

MODERN CERAMICS: SOME PRINCIPLES AND CONCEPTS. Edited by J. E. Hove and W. C. Riley. John Wiley & Sons, Inc., 1965.

THE PHYSICS AND CHEMISTRY OF CERAMICS. Edited by Cyrus Klingsberg. Gordon and Breach, 1963.

The nature of glasses

THE INITIAL STAGES OF PHASE SEPARATION IN GLASSES. J. W. Cahn and R. J. Charles in *Physics and Chemistry of Glasses*, Vol. 6, No. 5, pages 181-191; October, 1965.

PHYSICAL CHEMISTRY OF METALS. Lawrence S. Darken, Robert W. Gurry and Michael B. Bever, McGraw-Hill Book Company, Inc., 1953.

TECHNICAL GLASSES. M. B. Volf. Sir Isaac Pitman and Sons, Ltd., 1961.
THERMODYNAMICS OF SOLIDS. Richard A. Swalin. John Wiley & Sons, Inc., 1962.

The nature of polymeric materials

FIBRES FROM SYNTHETIC POLYMERS. Edited by Rowland Hill. Elsevier Publishing Company, 1953.

HIGH POLYMERS: A SERIES OF MONOGRAPHS ON THE CHEMISTRY, PHYSICS AND TECHNOLOGY OF HIGH POLYMERIC SUBSTANCES. Edited by R. E. Burk, H. Mark, E. O. Kraemer and G. S. Whitby. Interscience Publishers, Inc., 1941.

LINEAR AND STEREOREGULAR ADDITION POLYMERS. N. G. Gaylord and H. F. Mark. Interscience Publishers, Inc., 1959.

MECHANICAL PROPERTIES OF POLYMERS. Lawrence E. Nielson. Reinhold Publishing Corporation, 1962.

NATURAL AND SYNTHETIC HIGH POLYMERS. Kurt H. Meyer. Interscience Publishers, Inc., 1942.

POLYMERS AND RESINS: THEIR CHEMISTRY AND CHEMICAL ENGINEERING. Brage Golding. D. Van Nostrand Company, Inc., 1959.

PRINCIPLES OF POLYMER CHEMISTRY. Paul J. Flory. Cornell University Press, 1953.

TEXTBOOK OF POLYMER CHEMISTRY. Fred W. Billmeyer, Jr. Interscience Publishers, Inc., 1957.

The nature of composite materials

CERAMIC AND GRAPHITE FIBERS AND WHISKERS: A SURVEY OF THE TECHNOLOGY. L. R. McCreight, H. W. Rauch, Sr., and W. H. Sutton, Academic Press, 1965.

STRONG SOLIDS. Anthony Kelly. Oxford University Press, 1966.

The thermal properties of materials

ELECTRONS AND PHONONS. J. M. Ziman. Oxford University Press, 1960.

QUANTUM THEORY OF SOLIDS. R. E. Peierls. Oxford University Press, 1955.

The electrical properties of materials

ORGANIC MATERIALS IN ELECTRONICS. R. R. Neiman and R. E. Johnson in *Modern Science and Technology*, edited by Robert Colborn. D. Van Nostrand Company, Inc., 1965.

SEMICONDUCTORS. N. B. Hannay. Reinhold Publishing Corporation, 1959.

THEORY OF THE ELECTRICAL PROPERTIES OF GERMANIUM AND SILICON.
Harvey Brooks in *Advances in Electron Physics: Vol. VII*. Academic Press, 1955.

The chemical properties of materials
CHEMICAL PHYSICS OF SEMICONDUCTORS. J. P. Suchet. D. Van Nostrand Company, Ltd., 1965.
THE CHEMISTRY OF IMPERFECT CRYSTALS. F. A. Kroger. North-Holland Publishing Company, 1964.
SOLID-STATE CHEMISTRY. N. B. Hannay. Prentice-Hall, Inc., 1967.

The magnetic properties of materials
FERROMAGNETIC DOMAINS: A BASIC APPROACH TO THE STUDY OF MAGNETISM. E. A. Nesbitt. Bell Telephone Laboratories, 1962.
MAGNETISM. Edited by George T. Rado and Harry Suhl. Academic Press, 1966.
PHYSICS OF MAGNETISM. Soshim Chikazumi. John Wiley & Sons, Inc., 1964.

The optical properties of materials
THE INTERACTION OF LIGHT WITH LIGHT. J. A. Giordmaine in *Scientific American*, Vol. 210, No. 4, pages 38-49; April, 1964.
THE MODERN THEORY OF SOLIDS. Frederick Seitz. McGraw-Hill Book Company 1940.
OPTICAL MASER OSCILLATION IN A GASEOUS DISCHARGE. A. Javan in *Advances in Quantum Electronics*, edited by Jay R. Singer, Columbia University Press, 1961.
QUANTUM ELECTRONICS AND COHERENT LIGHT. Edited by P. A. Miles, Academic Press, 1964.

The competition of materials
METALS HANDBOOK. Taylor Lyman. American Society for Metals, 1961.
RIVALRIES BETWEEN METALS WITH OTHER MATERIALS IN METALLURGICAL ACHIEVEMENTS. W. O. Alexander. Pergamon Press, 1965.
THE SIMPLE ECONOMICS OF RIVALRIES BETWEEN MATERIALS. W. O. Alexander in *Contemporary Physics*, Vol. 8, No. 1, pages 5-20; January, 1967.